Lubrication in Practice

MECHANICAL ENGINEERING

A Series of Textbooks and Reference Books

EDITORS

L. L. FAULKNER
Department of Mechanical Engineering
The Ohio State University
Columbus, Ohio

S. B. MENKES
Department of Mechanical Engineering
The City College of the
City University of New York
New York, New York

OTHER VOLUMES IN PREPARATION

Lubrication in Practice

Edited by
W. S. Robertson

Second edition

MARCEL DEKKER
New York and Basel

Library of Congress Cataloging in Publication Data
Main entry under title:

Lubrication in practice.

   Includes index.
   1. Lubrication and lubricants.   I. Robertson, W. S.
TJ1075.L79   1984   621.8′9   83–20974
ISBN 0–8247–7204–0

Distributed in the Western Hemisphere by Marcel Dekker, Inc.

Printed in Great Britain

COPYRIGHT © 1984 by Esso Petroleum Company Limited
ALL RIGHTS RESERVED

Neither this book nor any part may be reproduced or transmitted in any form or by
any means, electronic or mechanical, including photocopying, microfilming, and
recording, or by any information storage and retrieval system, without permission
in writing from the publisher.

MACMILLAN PRESS LTD
Houndmills, Basingstoke, United Kingdom

Current printing (last digit):
10 9 8 7 6 5 4 3 2 1

Contents

Preface

Lubrication is important to every student, engineer and manager in industry: to help keep machines or production lines operating, to improve metal-working productivity, and as a vital part of industrial design.

Many books are available that comprehensively cover the theory of lubrication and other aspects of tribology. However, engineers and managers often need to know what lubricants do, and how they can best be used, rather than precisely how they function. That is the approach that *Lubrication in Practice* has taken. Its authors have all worked, in Esso and in other companies, on a wide range of practical lubrication and lubricant applications.

The first chapter summarises basic lubrication theory but does not attempt to deal with it in any detail. Chapter 2 discusses the types and properties of lubricants. The next eleven chapters cover such specific applications as diesel and petrol engines, hydraulics, compressors, machine tools and cutting oils. The final two chapters are on the storage and handling of lubricants, and on lubrication planning.

The information in *Lubrication in Practice* is designed to help people in industry increase overall efficiency and save money by improving lubrication practices. However, this information is necessarily general in scope and cannot cover all circumstances. It is therefore always desirable to take specific advice, whether from consultants, specialists or qualified lubricant suppliers, in applying this information to individual cases.

The second edition of this book has been completely updated in the light of developments since the first edition was published in 1972. This edition is being published in the centenary year of Beauchamp Tower's first report to the Institution of Mechanical Engineers on the Friction of Lubricated Bearings. Wider knowledge of the practical implications of such studies is quite as important now as it was then.

The potential financial benefits are much more important now. Nearly 20 years ago, in 1966, the Jost Report said that Britain could save £500 million a year by better tribological practices, including improved lubrication. That saving is equivalent to at least £2000 million a year today.

Lubrication in Practice is designed to give its readers some practical assistance in their efforts to achieve, within their own organisations, some share of that potential saving.

1 Theory and Basic Principles

R. T. Davies *B.Sc., A.R.I.C.*
A. J. S. Baker *C.Eng., M.I.Mech.E., A.M.I.Mar.E., A.F.Inst.Pet.*

The basic purpose of lubrication is the minimisation of friction and wear. Normally, the lubricant is required to perform various auxiliary duties, cooling or cleansing for example, but these secondary roles will be ignored in this chapter.

Friction is defined as the force resisting motion when two contacting surfaces are moved relative to each other. In a dry system, this force is a result of interaction between the surfaces involved; a fundamental difference exists in the nature of this interaction depending on whether the motion is of a sliding or a rolling character. In a lubricated system, these mechanisms will be weakened to the extent that a coherent fluid film is interposed between the surfaces and an additional but lesser force dependent on the viscosity of the fluid will be present.

Where complete separation is attained, the lubricant viscosity becomes the controlling parameter, and the condition is termed *hydrodynamic lubrication.* Where surface interaction continues to exert a significant effect, the term *boundary lubrication* is used.

In the absence of definite contamination, wear is the result of friction and in a simple case is roughly proportional to it. Again, different mechanisms exist; differentiation should be made between the roles of adhesion, abrasion, corrosion and fatigue, even though the final damage may be the end-product of a combination of these mechanisms.

The ensuing sections of this chapter expand on this generalised summary of basic features and examine the function of the lubricant in greater detail.

NATURE OF SURFACES AND DRY FRICTION

An understanding of the fundamental mechanism of friction presumes an appreciation of the nature of nominally flat surfaces on a microscopic scale. This shows a finely polished finish as a series of hills (asperities) and valleys (troughs) of finite dimensions, while normal engineering finishes are viciously contoured. Hence, if one such surface is brought in contact with another the actual contacting points are few and correspond to the limited number of high asperities on one surface

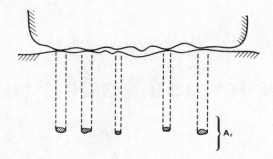

Figure 1.1   Contact areas of 'welded' junctions

that meet corresponding points on the other. These few junctions must bear the total load, and pressures at these points are very high, causing virtual welding (adhesion) of one point to another. The force of sliding friction, therefore, equates to the force required to shear these junctions, the strength of which are dependent on the load but not on the apparent total area of contact. For chemically clean surfaces this force is very high, but in the normal practical case it is considerably weakened by contaminant films of oxide, moisture and occluded gas which are present on any normal surfaces. Where materials of dissimilar strength are slid together, a secondary term must be included to cover the additional force required to plough the asperities of the harder material through the softer one.

N.B.

Therefore, in its simplest form the equation governing sliding friction (see figure 1.1) can be shown as

$$F = A_r S + p$$

where A_r = real contact area of 'welded' junctions (dependent on load and yield characteristics of materials),

   S = shear strength of junctions,

   p = force required to plough the harder asperities through the softer material (again dependent on load and yield characteristics).

The above theoretical derivation of the force that resists motion when one surface is slid over another cannot be applied to the case where one surface is rolled over another except insofar as micro-slip occurs. In pure rolling, the greater part of the resistance, which is of considerably lower order of magnitude, is due to hysteresis losses in the materials themselves and sub-surface plastic deformation under conditions of high stress. In the practical case, ball bearings for example, the simplicity of this concept is marred by such confusing factors as skidding and cage restriction which add complexity to the problem and partially re-insert the features attributed to sliding friction.

General lubrication

From the foregoing considerations, it will be appreciated that the prime purpose in lubrication is to interpose between the surfaces a quantity of low-shear material that is adequate to prevent or at least minimise asperity contact. The extent to which this can be achieved by engineering design, either throughout or partially during normal service conditions, is an important yard-stick of success.

HYDRODYNAMIC LUBRICATION

Hydrodynamic lubrication is the term used to cover the regime where complete separation of the moving surfaces has been attained throughout the majority of service. Its achievement depends on geometrical design and maintenance of adequate speed of movement.

The establishment of a lubricant film of satisfactory thickness is dependent on several conditions from which the load-carrying capacity of the film and hence the ability of the surface to remain separated under load can be established.

Qualitatively the controlling factors for full-film condition are as follows:

Attitude	Surfaces must adopt a very slight taper or wedge attitude to generate a film pressure
Sliding velocity	Needed to drag sufficient oil into the contact by viscous resistance
Oil supply	Minimum quantity of oil needed to fill the clearance and make up losses from the edges of the surfaces
Viscosity	A viscous attraction of the oil to the surfaces and flow determines the thickness of film generated

These requirements are expressed mathematically in the classic Reynolds[1] equation which may be written in the general form that assumes an incompressible oil and two-dimensional conditions:

$$\frac{d}{dx}\left(\frac{h^3}{12\eta}\frac{dp}{dx}\right) = \frac{d}{dx}\left(\frac{u_1 + u_2}{2} \cdot h\right) + \left(w_2 - w_1 + u_1\frac{dz_1}{dx} - u_2\frac{dz_2}{dx}\right)$$

where
h = oil film thickness,
u = sliding velocity,
w = load/unit width,
p = instantaneous oil pressure,
η = absolute viscosity,
x = linear parameter,
z = radial parameter,
suffixes 1 and 2 being used to denote maxima and minima in the system.

Plain journal bearings

The above conditions are most easily satisfied in the plain journal bearing having a continuous oil supply at mid-length and a small radial clearance between the journal and bearing.

Figure 1.2 shows a bearing at its three stages of full-film separation. When starting from rest at (a) the journal rests on the bottom of the bearing with the surfaces in contact; as it starts to rotate a slim wedge of oil is dragged between the surfaces and they separate (b). As the journal speed increases the film thickness increases to further displace the journal to its stable running position (c). From this the importance of tapering film thickness afforded by the bearing clearance will be noted. In practice this clearance will, of course, be very small — of the order of 1×10^{-3} of the journal diameter.

In this simple case, the bearing has been considered as being full of oil before start up. However in most practical cases oil is supplied by a pump connected to

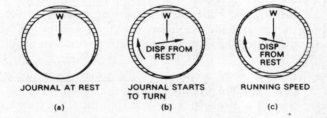

Figure 1.2   Plain journal bearing at its three stages of full-film separation

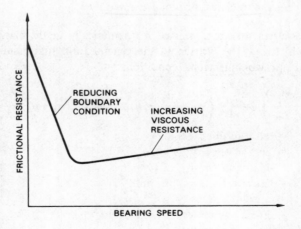

Figure 1.3   The transition from boundary to hydrodynamic lubrication

the same shaft so that the bearing must rotate for the first few revolutions without additional oil. During this period the lubrication depends on residual oil left on the surfaces and it will be seen that the conditions must be those of boundary lubrication, mentioned earlier, which accounts for the care taken in practice to ensure compatibility of bearing and journal materials.

The transition from boundary to hydrodynamic lubrication is accompanied by marked changes in the frictional resistance of a bearing as it speeds up from rest, as shown in figure 1.3. The initial steep reduction in resistance is accounted for by the transition from boundary to full-film lubrication. It will be noted that the lowest level of frictional resistance is obtained with just sufficient speed to separate the surfaces. Increases of speed beyond this point give rise to greater resistance as a result of the additional work being done in shearing the oil film.

Viscous resistance

The effort needed to shear a film of oil is caused by the internal cohesiveness of the oil, known as its viscosity[2]. The same property coupled with interfacial attraction determines the thickness of the film attached to the surface to enable the oil wedge to be established. From this it will be seen that sliding between two surfaces separated by an oil film produces a shear strain on the film, as shown in figure 1.4. Here abcd represents a section of an oil film between two flat surfaces. When the upper surface moves to the left the oil attached to the upper surface moves with the surface (ab to $a_1 b_1$) while that attached to the lower surface remains stationary. The work done in straining the oil is represented by its resistance to flow or viscosity.

These properties are also responsible for the ability of an oil film to form a load-carrying wedge. In figure 1.5, two flat surfaces are again shown, but this

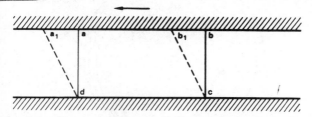

Figure 1.4   Shear strain on an oil film between two surfaces

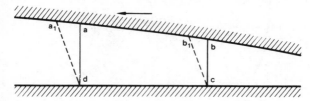

Figure 1.5   Shear strain on an oil film between two surfaces
that diverge at a small wedge angle

time they diverge at a small wedge angle. In practice, this angle occupies only a few minutes of arc. In this case the strained triangles aa_1d and bb_1c are similar but bb_1c is smaller. Assuming oil entering the wedge fills aa_1d, it must leave through the space bc at increased velocity from that of the entry conditions. The fluid resistance generated produces a pressure in the film, which tends to separate the surfaces, thus enabling them to carry a vertical load. It should also be noted that if the direction of sliding is reversed, the result will be a fall in pressure along the now reversed wedge and an absence of load-carrying ability. However, the low-pressure region thus created forms the basis for feeding oil to most cylindrical plain bearings.

The relationship of low-pressure and high-pressure regions[3] is best understood by considering the journal bearing as divergent and convergent flat surfaces bent into circular form as shown in figure 1.6. This also shows the region of low pressure in the oil film that must be taken into account in feeding oil into the bearing. To attempt this in the high-pressure region would require an oil feed pressure of perhaps 700 bar, whereas in the low-pressure region the bearing is perfectly capable of lifting its own supply from an adjacent duct, thus greatly facilitating the design of feed systems.

Additional considerations

Three further points on plain bearings are worthy of comment. Firstly, it will be realised that the collective forces, load, wedge pressure and viscous resistance,

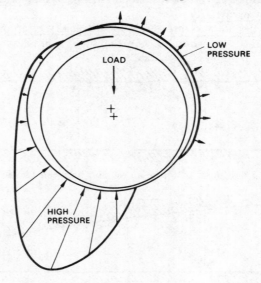

Figure 1.6 The relationship of high-pressure and
low-pressure regions in a journal bearing

must combine as a closed force polygon to maintain the journal in a stable position relative to the bearing. If this condition is not met, the journal centre will move in an effort to reach an equilibrium position. If equilibrium cannot be reached, the journal centre will continually change, usually at one-half the rotational speed of the shaft, to produce the phenomenon known as half-speed whirl[4]. However, happily this condition is seldom experienced in moderately loaded bearings at surface velocities of less than 3000 m/min.

Secondly, work is done in the oil when being sheared during each revolution of the bearing, which results in heat being generated. This heat is carried off in the oil but obviously results in a rise of temperature which, in turn, reduces the oil viscosity during its passage through the bearing. Thus, in preparing a bearing design due account must be paid to the reduction of viscosity[5,6].

Thirdly, shaft loads are seldom completely steady and may fluctuate quite widely during each rotational cycle. The sudden application of load results in the rapid displacement of the oil film and the generation of additional pressure in the film as a result of fluid friction. This is known as the squeeze[7] effect and provided the duration of the impulsive load is short it can prevent the collapse of the film. The importance of this factor will be realised from the fact that internal combustion engine big-end bearings are entirely dependent on squeeze effect to resist the shock of combustion without metallic contact in the bearing.

ELASTO-HYDRODYNAMIC LUBRICATION

The foregoing sections have covered what may be termed the classical hydrodynamic case, the chief characteristics of which can be summarised as very low friction and wear and dependence on a simple function of viscosity. This is distinct from boundary lubrication where some asperity contact inevitably occurs and friction and wear are significantly greater: lubricant viscosity alone appears to have little effect in this case.

It has long been recognised that an intermediate zone existed, particularly in the case of gears, that gave relatively low friction and wear and was viscosity-dependent, but which could not be supported theoretically using the classical hydrodynamic derivation of oil film thickness. A series of experimental studies in the early 1950s explored this area in detail and founded the extension of hydrodynamic theory known as elasto-hydrodynamics to explain this phenomenon.

Basically this approach recognises, and uses as additional terms, two further properties of the system. These are:

(1) The surfaces of the materials in contact deform *elastically* under pressure and hence the load is spread over a greater area.
(2) The viscosity of the lubricant increases dramatically at high pressure thereby increasing the load-carrying ability in the contact zone.

It will be appreciated that both these effects are reversible and only obtain at the time of contact.

A simple indication of the difference of magnitude in oil film thickness derived by classical and elasto-hydrodynamic theory is shown in figure 1.7 and in the equations for film thickness at the pitch line of a spur gear combination.

The representation on the left is the pure hydrodynamic one with

h = film thickness
R = relative radius of curvature
η_o = lubricant viscosity at entry to the contact zone
U = relative speed of approach
W = load.

Elasto-hydrodynamics postulates the same situation with an increased film thickness spread over a greater contact length and here

h_1 = film thickness
R, W and η_o have same significance as previously
E = elastic modulus of surface
α = viscosity–pressure coefficient.

It will be noted that the effect of load, in particular, is reduced considerably relative to its former value.

This increased understanding is a considerable advance over previous theory, and it is probable that future extension will enable even greater 'hydrodynamic' explanation for behaviour in the thin film zone. However, ultimately the limits are undoubtedly reached where physical explanation is no longer possible, as conventionally measured viscosity becomes a non-determinant, and chemical characteristics are seen to be overriding. This is the area of true boundary lubrication.

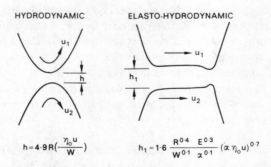

HYDRODYNAMIC   ELASTO-HYDRODYNAMIC

$$h = 4.9 R \left(\frac{\eta_o u}{W} \right)$$

$$h_1 = 1.6 \frac{R^{0.4} \; E^{0.3}}{W^{0.1} \; \alpha^{0.1}} \left(\alpha \, \eta_o u \right)^{0.7}$$

Figure 1.7   Oil film thickness derived by hydrodynamic and
elasto-hydrodynamic theory at the pitch line of a spur gear combination

BOUNDARY LUBRICATION

The physicochemical mechanisms that operate in boundary lubrication are still incompletely understood and are the subject of much current academic and commercial study. It is virtually certain that there is no single explanation and that different mechanisms and reactions obtain in particular applications, depending on the nature of the materials involved, their surface finish and the conditions. The regime is extremely responsive to additives in the basic lubricant and selection of the additive used is determinant. Whatever mechanisms are followed, however, the net effect is a limitation of intimate asperity contact by very thin films of the basic lubricant, or additive, or relatively low shear strength reaction products formed in the critical areas. Examples of the type of compounds that are effective in promoting this condition are discussed below.

Contamination and chemically prepared surfaces

As mentioned earlier, virtually all surfaces are contaminated with atmospheric gases, trace films of moisture, and 'oxide'-type reaction products. As such, coefficients of friction are much lower than those of the same surfaces when freshly exposed or after rigorous 'out-gassing' *in vacuo*. It will thus be seen that the contamination is operating as a natural boundary lubricant. Its magnitude is inadequate for the majority of engineering purposes but the parallel of deliberately 'reacting' a surface to give a low shear strength solid film is employed in various cases such as hypoid gears. Phosphorus and sulphur compounds are generally employed as an end-treatment in manufacture and act as an extra safeguard against rupture of normal liquid lubricant film during running-in.

Dispersion of solid lubricants

Two notable solids are used in this form, graphite and molybdenum disulphide. Both have a lamellar crystal structure which is built of series of flat platelets lying one above another. This structure is strong in compressive but weak in shear strength and sliding spreads the platelets over the surfaces involved, minimising true contact.

Long-chain polar compounds (and weak acids)

It has long been recognised that natural fats and oils have superior lubricity particularly at low sliding speeds and within a limited temperature range. This behaviour is due to molecular structure wherein a long hydrocarbon chain is attached to a polar end group (such as long-chain alcohols, esters and acids). At a normal lubricant/surface interface these molecules orient themselves with polar group to the metal and chain to the oil in a close-packed formation. The result is analogous to the pile on a carpet and provides an adherent low shear strength film on the surfaces involved. The same effect can be produced by adding long-chain polar compounds to mineral oil.

The major drawback with these systems is ultimate lack of strength and extreme temperature sensitivity, since increase of temperature promotes desorption. Long-chain acids have the greatest utility, as in this case simple physical attraction is reinforced by chemical reaction to form a boundary film of salts (or soaps) formed between the metal and the acid. This film breaks only when the melting point of the soap is reached.

Reactive compounds

The majority of commercial EP (extreme pressure) and anti-wear additives fall into this classification. They are predominantly organic compounds containing sulphur, phosphorus or chlorine. In the crudest case (for example, when flowers of sulphur is added to mineral oil), their action can be seen as overall reaction with the surfaces involved to form a low-strength chemical film (for example, iron sulphide) which protects the true surface when the fluid lubricant film is ruptured. This is analogous to the pre-chemical treatment of surfaces mentioned earlier. The disadvantage of this approach is its uncontrolled nature with subsequent problems of corrosion.

The absolute mode of action of the more sophisticated organic compounds is still contentious but the simplest generalisation is that they react only at points of incipient scuffing, their reactivity being promoted by the rise in temperature at these points and the extreme surface energy of any freshly exposed metal. The product of this reaction can be viewed as a low shear strength slurry that minimises further contact.

Friction polymer

Finally, the latest theories on friction polymers must be mentioned. This approach recognises the presence of certain molecular species in normal lubricating oils that are capable of forming high-viscosity/solid polymer under the same influences as those causing the EP reactions mentioned above.

The authors recognise that, throughout this section on boundary lubrication, they have grossly simplified mechanisms in the interest of limiting space and ease of reading. The reader is referred to the bibliography if he wishes to pursue the subject to greater depth.

WEAR

The postulation of definitive laws of wear is considerably more recent than the corresponding development for friction. Wear behaviour is much more variable and is affected by a wider variety of determinant causes: frequently these causes interlock to give a cumulative result.

Four major mechanisms can be established. These are: adhesion, abrasion, corrosion, and fatigue. Adhesive wear is the primary frictional process arising from surface rupture at asperity welds. Abrasion is the ploughing process arising when a hard asperity or detached fragment runs over a softer material. Corrosive wear is dependent on the atmosphere under which movement occurs: in the static state, a coherent and possibly protective film of reaction product would form, but under a sliding condition this is continuously removed leaving a naked surface susceptible to further reaction. Fatigue wear is a sub-surface phenomenon, particularly associated with rolling contact, resulting in eventual collapse of the surface as a result of the weakening of its foundation.

In this simplified analysis other nominated forms of wear are seen as sub-categories of the four major ones listed. Thus, fretting is a combination of adhesion, abrasion and corrosion; erosion is a form of abrasion; and cavitation is a form of fatigue.

It will be appreciated that with the exception of gross cases of corrosion, erosion and cavitation, wear processes are limited to dry and boundary conditions, while the hydrodynamic case is absolved.

In practical engineering terms it is extremely difficult to separate pure adhesion and abrasion in cases where surfaces of dissimilar hardness are employed, and in any case the latter must quickly follow the former where there is repeated traverse. A pure theoretical approach postulates adhesive wear (volume worn away) as a function of the form:

$$\frac{K L X}{p}$$

where K is a constant for the materials in contact,
       L is the load,
       X is the distance slid,
       p is the flow pressure of the softer material

while abrasive wear inserts an additional term, dependent on the geometry of the average asperity on the harder surface, into the numerator.

The role of the lubricant in this complex system is identical with that discussed for the minimisation of friction by limiting asperity contact either physically or by chemical modification to a less injurious nature. A secondary, but no less important duty is the removal of loose fragments from the contact area to a less critical zone where they can be filtered out of the oil.

Corrosive wear is obviously dependent on the atmosphere obtaining in the surface area during both the operating cycle and shut-down periods. This atmosphere can vary from a normal mildly oxidative system, such as dry air, to an intensely corrosive liquid/gas-phase interface containing sulphur oxides, HCl, HBr,

etc. In the case of the former condition the lubricant's role is passive and maintenance of a coherent film of even monomolecular dimensions can diminish simple
oxidative corrosion. In the latter case, the lubricant can assume a very active role
in containing controlled alkalinity that is capable of neutralising the acid present
to form inactive salts which are then carried away from the critical area.

Fatigue wear, arising from sub-surface damage as a result of repeated stress
cycling, is most frequently encountered in rolling element bearings and results
in a sudden, and often unpredictable, rapid pitting and flaking of the surface. It
is not at all clear whether improvements in fatigue life can be obtained by lubricant modifications (assuming that the lubricant is already ideal in controlling
other forms of wear present in the application). It is known that highly surface-
active additives have a harmful effect but it does not appear that much advance
has been made in a positive direction.

As stated in the pre-amble, the eventual manifestations of wear often present
a confused picture resulting from different mechanisms occurring concurrently
or being overlaid one upon another. In any analysis of a particular wear case, it
is essential to consider all aspects throughout its history and attempt to recon-
struct the chain of events so that the causative stage may be highlighted.

CONCLUSION

In writing this introductory chapter the authors have been very conscious of the
fact that the breadth of the field to be covered within a limited space has necessi-
tated an extremely cursory treatment. Also, the major accepted difficulty in the
subject is its interdisciplinary nature and, in an effort to simplify it for readers of
individual disciplines, several broad generalisations have been made that must
appear contentious to the specialist.

We make no apology for this, as the primary objective has been to construct a
platform from which the reader can better approach the later specific chapters
and turn to specialist literature in his particular field of interest.

The Bibliography contains several of the leading text books in the field and
the reader is recommended to consult them, not only for their textual content
but also as sources of reference to individual work.

In addition, the technical departments of the major oil companies are well
equipped to give advice on their own particular products, while the National
Engineering Laboratory (East Kilbride) and Industrial Tribology Units (Leeds,
Swansea and Risley) have been set up as service units to industry in this field.

REFERENCES

[1] Osborne Reynolds, *Phil. Trans.*, (1886) 157.
[2] E. Hatschek, *The Viscosity of Liquids*, G. Bell, London, 1928.
[3] Beauchamp Tower, *Proc. Inst. Mech. Eng.*, (1883) 632; (1885) 58.
[4] F. R. Archibald, The stepped shape oil film applied to journal bearings, *J. Franklin Inst.*, 253 (January 1952) 21.
[5] R. A. Baudry, Some thermal effects in oil-ring journal bearings, *Trans. Am. Soc. Mech. Eng.*, 67 (1945) 117.
[6] D. Dowson *et al.*, An experimental investigation of the thermal equilibrium of steadily loaded journal bearings, *Bearings for Recip. and Turbo. Machinery Conference Proc., Institution of Mechanical Engineers, 1968*, p. 70.
[7] F. P. Bowden and D. Tabor, The wear and damage of metal surfaces with fluid lubrication, no lubrication and boundary lubrication, *Am. Soc. Metals*, (1950) 109.

BIBLIOGRAPHY

F. T. Barwell, *Lubrication of Bearings*, Butterworth, London, 1956.
A. Bondi, *Physical Chemistry of Lubricating Oils*, Reinhold, New York, 1951.
F. P. Bowden and D. Tabor, *The Friction and Lubrication of Solids. Parts 1 and 2*, Clarendon, Oxford, 1954 (Part 1), 1964 (Part 2).
A. Cameron, *Principles of Lubrication*, Longmans, London, 1966.
D. Dowson and G. R. Higginson, *Elastohydrodynamic Lubrication – in S.I. units*, Pergamon, Oxford, 1977.
J. Halling, *Principles of Tribology*, Macmillan, London, 1979.
E. Rabinowicz, *Friction and Wear of Materials*, Wiley, New York, 1965.

2 Types and Properties of Lubricants

W. S. Robertson *C.Eng., F.Inst.E.*
Industrial Communication Associates

This chapter outlines the main types of lubricant and covers some of the more important lubricant properties and tests, finishing with engine tests. With each test description there is a note on any relevant additives or classification systems.

TYPES OF LUBRICANT

Lubricants can be fluids (gases or liquids) or solids.

Fluids

Gases are not always considered as lubricants, but air-lubricated bearings are quite widely used for special purposes. However, air for specialist applications may not be a cheap lubricant. For hydrostatic spindle bearings on machine tools, for instance, air cleaned and dried to adequate standards can cost several times as much as lubricating oil.

Lubricating oil – hydrocarbon oil produced from crude petroleum – is of course the most common of all lubricants. Hydrocarbon oils are used because:

(1) They are available in a range of viscosities that gives a wide choice of load, speed and temperature conditions to the designer.
(2) They give a low, consistent coefficient of friction and have low compressibility.
(3) They are reasonably effective in carrying away heat from bearing surfaces.
(4) They are inexpensive lubricants.

Hydrocarbon lubricating oils fall into two main categories:

Paraffinic oils have high pour points (because of the wax they contain), high viscosity indices and good resistance to oxidation.

Naphthenic oils have low pour points and relatively low viscosity indices and oxidation stability.

The above properties are described later in this chapter.

Synthetic lubricants (esters, phosphates, silicones) may cost several times as much as hydrocarbon oils but they are necessary for some specialised applications, like aircraft gas turbines where resistance to degradation at temperatures of up to perhaps 300 °C is necessary.

Water is much better at carrying away heat than are hydrocarbon oils, but its low viscosity and load-bearing ability, and its relatively high freezing point and low boiling point, greatly limit its use as a lubricant.

Solids and semi-solids

Grease is the most important solid or semi-solid lubricant, and it is normally made from hydrocarbon oils thickened with metallic soaps, in consistencies ranging from slightly thickened liquid to block-hard (see Chapter 11). Although grease will not carry heat away from a bearing as liquid lubricant will, it can be an effective seal against dirt and water reaching the bearing surfaces. It can also provide a reservoir of lubricant in a bearing lubricated at long intervals, or even sealed for life.

Other solid lubricants, functioning in a different way, include molybdenum disulphide and graphite. These may be used on their own or in combination with oils and greases. Chemical coatings on bearing metals and bearings made from plastics such as PTFE can also be included in this category.

The remainder of this chapter is concerned only with conventional industrial lubricants made from hydrocarbon oils, either alone or with additives to enhance their natural properties.

PROPERTIES OF LUBRICANTS

The first, obvious, property of a lubricant is its ability to keep moving surfaces apart, in all the conditions of pressure and temperature, and in the presence of contaminants etc., to which they may be subjected.

Secondly, most lubricants also need to act as a cooling medium, by carrying away the heat generated at the bearing surfaces or reaching them from an outside source (for instance, in the way that the heat of combustion of a gas turbine reaches the turbine bearings).

Thirdly, a lubricant should be stable enough to keep on doing its job for its designed life-time, whether that is a few seconds on a once-through mist-lubricated bearing or ten or more years in a steam turbine. Lastly, the lubricant should protect the surfaces with which it comes into contact against corrosion from the atmosphere or from corrosive products generated in the equipment, such as acid gases from an internal combustion engine.

TESTS ON LUBRICANTS

Purpose of tests

Measurement of the properties of lubricants can show lubricant composition and can indicate, to some extent, lubricant performance, but hydrocarbon oils are too complex for this type of testing to be a complete guide to behaviour in practice. Laboratory engine tests and, finally, controlled field trials in the actual equipment to be lubricated are needed to prove the suitability of a particular oil for a particular use.

The chemical composition of the oil makes a great difference to lubricant performance, and the additives used with the oils also considerably affect performance. However, tests that indicate the presence of particular types of oil or additive still need to be supplemented by performance tests. Tests for composition are really most useful for process control in ensuring that, once a specification has been established for a lubricant, the lubricant is always produced with as near the same composition as is necessary.

Tests and specifications

There are two kinds of specifications for lubricants: the manufacturing specifications just mentioned above, set up by makers to keep their products consistent in composition and therefore in quality; and performance specifications set up by standards institutions like the British Standards Institution, by equipment makers like Caterpillar or Ford, and by governments, primarily for military use.

Performance specifications range from the Society of Automotive Engineers crankcase oil classifications, which specify only a range of viscosities and related qualities, to very detailed military specifications like MIL-L-2104D for engine oils.

Specifications of this last kind are built up from standard laboratory and engine tests. Tests on lubricants have to be carefully standardised because few of the tests measure fundamental properties that are independent of the test method. If the method or the equipment used for the test is changed, the result will change too.

Specifications will always give either an upper or lower limit or a range of properties for each test result, because no production lubricant can be made to give exactly the same results every time, nor are the tests used capable of exact repeatability every time — there must be a range within which the results can fall.

SPECIFIC TESTS ON LUBRICANTS

Various standard tests are mentioned in this section. None are described in detail because full technical information on all of them can be found in the relevant Institute of Petroleum[1] and the American Society for Testing and Materials publications[2], and in other publications[3,4] that describe the significance of tests more fully than is possible in this chapter.

Viscosity

Viscosity and lubrication theory has been covered in Chapter 1. Some basic points are repeated here to lead into a description of viscosity test methods.

Viscosity, a measure of resistance to flow in a fluid, is the most important of all lubricant properties. Viscosity increases with increasing molecular weight and changes rapidly with changes in temperature. Heating a lubricating oil lowers its viscosity; cooling raises it.

Design calculations require a knowledge of the dynamic viscosity. In streamline flow conditions, dynamic viscosity is defined as the tangential force on unit area of one of two planes parallel to each other, separated by unit distance, and moving with uniform unit velocity relative to each other, the space between being filled with fluid. The SI unit is $N s/m^2$ but the centipoise ($10^{-3} N s/m^2$) is normally used.

Kinematic viscosity (see figure 2.1) is universally used for lubricant viscosity

Figure 2.1 Automatic kinematic viscometer – this instrument automatically measures viscosity with an accuracy and reproducibility at least as good as that of manual viscometers

measurement in the oil industry. It is the ratio of the dynamic viscosity to the density of the fluid at the same temperature. Kinematic viscosity is readily determined in suspended level glass viscometers in constant-temperature baths. The SI unit is m^2/s but the centistoke (cSt), which is 10^{-6} m^2/s, is normally used.

Viscosity and temperature

Oils vary in viscosity with temperature, and different oils vary by different amounts for the same temperature change. The less this change the more effective the oil to lubricate in applications like an internal combustion engine where it may start cold and run hot.

The most widely used measure of relative-viscosity change with temperature is viscosity index (VI). An oil with a low viscosity index changes greatly in viscosity with change in temperature; an oil with high viscosity index changes relatively little in viscosity for the same temperature change.

When the system was set up, oils were compared with two reference oils arbitrarily given the viscosity indices of 0 (worst) and 100 (best), but many oils now have viscosity indices above 100.

A less direct way of indicating good viscosity/temperature behaviour is the 'multigrade' classification for engine oils. This is based on the Society of Automotive Engineers (SAE) crankcase oil classification (see Chapter 3). It happens that oils with high VIs can meet the requirements of more than one grade simultaneously, and the 'multigrade' oils, say SAE 10W/30 or SAE 10W/50, will necessarily have good viscosity/temperature characteristics.

Viscosity at low temperature

Viscosity at low temperature is important for ease of starting in the cold in, for example, diesel or petrol engines. It can be calculated from standard equations and charts if the viscosity at two higher temperatures is known, but this method does not give accurate results. It can be determined directly in glass viscometers in low-temperature baths, but this method too does not always relate to performance. The reason is that lubricating oils at low temperature, and particularly when they contain polymeric additives, have non-Newtonian properties (apparent viscosity varies with rate of shear) and their viscosity cannot be predicted by methods that depend on extrapolation of Newtonian properties determined at higher temperatures.

Cold cranking performance in engines can be found by measuring the force needed to turn the actual engine at low temperature. This is accurate, but expensive.

A cold cranking simulator test has been developed for laboratory use (see figure 2.2). It is simple and rapid to perform and correlates well with engine performance, and it is now required in the SAE classification.

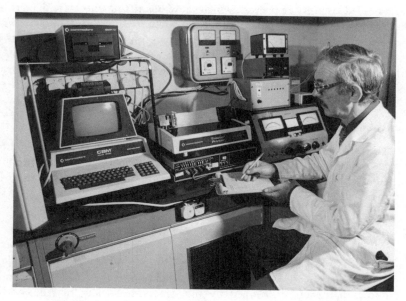

Figure 2.2   Automotive engine cold cranking simulator

As part of the SAE classifications mentioned earlier, a mini-rotary viscometer is used to measure viscosities at temperatures between -15 °C and -38 °C. These are graphed to give the border-line pumping temperatures that are now included in the classification.

Viscosity and volatility

Viscosity gives an indication of the volatility of a lubricant: in general, the lower its viscosity the higher its volatility. Viscosity therefore correlates to some extent with engine oil consumption since, other things being equal, consumption is affected by oil volatility.

Viscosity only indicates volatility. (Flash point is another indicator, at least to the extent of showing contamination by low-volatility material like petrol or diesel fuel.) To obtain an accurate picture of lubricant volatility, a distillation/temperature graph would be required.

Viscosity improvers

Selection of oil base stock and oil-refining methods can give lubricants of quite high VI, but even better viscosity/temperature performances are often needed to meet requirements for, say, engine oils or hydraulic oils. This can be achieved by additives called viscosity improvers, which cause oils to thin out less with increase in temperature than they would without the additive. Viscosity improvers have to be selected carefully and matched to the base oil and other additives that may be used. They vary in the extent to which they remain effective in use. A poorly chosen viscosity improver breaks down in service, causing the oil to drop in viscosity as well as in viscosity index.

Viscosity and friction modification

Although viscosity is the most important lubricant property governing friction in engine lubrication, it is not the only one. Lubricants can also be friction-modified by suitable additives to give lower friction at the same viscosity as before. This can lead to lower engine fuel consumption.

Viscosity classifications

SAE crankcase oil classification

One viscosity classification has already been mentioned, the SAE classification of crankcase lubricants, developed by the American Society of Automotive Engineers[4]. It is given in full in Chapter 3.

There is a similar SAE viscosity classification for gear lubricants.

Viscosity classifications for industrial liquid lubricants

The British Standards Institution has a standard (BS 4231) for viscosity classification of liquid lubricants[5].

ISO and BSI machine tool lubricants classification

ISO 3498-1979(E) and BS 5063: 1982 cover lubricants for machine tool use (see Chapter 9).

Extreme pressure and anti-wear properties

In extreme conditions, hydrodynamic lubrication cannot be maintained and extreme pressure and anti-wear properties will become more important than viscosity. Examples are in gears, particularly hypoid gears, and in some high-speed engine parts.

Several different rig tests are used to measure extreme pressure properties, including the SAE, Timken, Falex, and Four Ball machines, and there are also numerous tests using standard production gear and bearing assemblies. Some standard engine tests include assessments of wear, which can be related to load-carrying ability.

Extreme pressure and anti-wear additives

There is a range of additives with increasing 'oiliness' or extreme pressure properties, based on fatty oil blends for increased oiliness, and on phosphorus, lead, chlorine or sulphur derivatives, or combinations of these for extreme pressure conditions. They act in extreme pressure conditions by forming compounds of the metal of contacting surfaces under the very high-temperature conditions generated locally at points of contact. These compounds prevent welding of the metal surfaces at points of contact, and so reduce friction between the surfaces. Extreme pressure conditions are dealt with in Chapters 1 and 6.

Oxidation stability

All hydrocarbon oils react with oxygen in the air, eventually forming acid or sludge products. These products could cause bearing corrosion or blocking of oil lines or filters. The time taken for oil to oxidise depends on operating conditions; an oxidation-inhibited turbine oil may stay in service for ten years but a diesel engine oil will not last in the high temperature and contaminated environment for more than a few thousand hours. There are numerous laboratory tests for oxidation stability including one for transformer oil and one for turbine oils. All the tests depend on artificially accelerated oxidation, by heating the oil and by blowing air or oxygen through it, in some cases in the presence of a catalyst (see figure 2.3). Even accelerated oxidation tests may take several thousand hours to complete on inhibited oils because of the high resistance to oxidation that such oils have.

Oxidation inhibitor additives

Oxidation inhibitor additives may operate at relatively low temperatures, as in steam turbines, or at high temperatures, as in diesel engines. Both types of additive are used in small amounts and operate by inhibiting the formation of intermediate oxidation products or by preventing metals present in the system from catalysing the oxidation reaction. Zinc dialkyldithiophosphate (ZDDP) is both an oxidation inhibitor and a film-strength (wear-resisting) additive, used principally in hydraulic and engine oils. Turbine oil inhibitors are usually amine or phenol derivatives.

Corrosion prevention

Moisture present in the oil or condensing from the atmosphere can cause corrosion of metals in engines and circulating systems. Rust inhibitors in the oil protect against this type of corrosion.

Engine bearings can be corroded by acids formed from combustion products and from the products of oil degradation. Oxidation inhibitors prevent acids

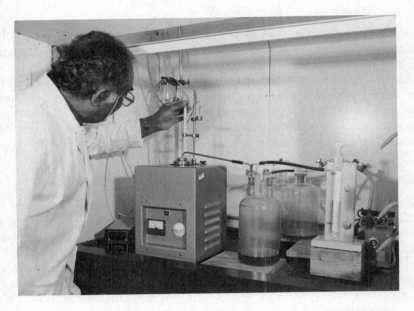

Figure 2.3   The NOAK oxidation test, based on a German DIN
specification, is now widely used in Europe

forming from oil oxidation, and many detergent-dispersant additives have
neutralising properties, so preventing acid corrosion.

There is a laboratory test for corrosion prevention of oils in circulatory
systems. Corrosion prevention of engine bearings is measured by a standard
engine test.

Rust inhibitors

Rust inhibitors form a thin protective film on metals when used in a low concentration in turbine or similar oils. They are polar organic compounds.

Pour point

Because mineral oils are complex mixtures, they have no sharp freezing point
like, say, water. The temperature at which an oil stops flowing in given conditions is governed by the conditions of the test. The rate of cooling, the shape
of the container holding the oil, and the amount that the container is moved, are
all important factors.

Different types of oils have widely different pour points. Those derived from
paraffinic bases have a large amount of waxy components which make the oil

Figure 2.4   Autopour equipment eliminates the need for frequent
checking in pour point determinations

stop flowing at a much higher temperature than oils derived from naphthenic bases, even if the naphthenic base oils have the same viscosity as the paraffinic base ones. Many finished oils are blends of both types of base oils and their pour points, and the pour points of purely paraffinic base oils, can be reduced by removing the waxy components from the oil. However, de-waxing is expensive. Another effective method of reducing pour points is to use additives — pour point depressants — to supplement a moderate amount of de-waxing.

Pour point depressants modify the structure of waxy oils so that, as the oil cools, the wax does not form a structure that would otherwise trap the rest of the oil and so prevent flow or block filters.

A standard pour point test consists of cooling oil at a standard rate until it just fails to flow. An automated version of this test is now widely used (see figure 2.4). More complex tests show ability of an oil to flow under applied pressure, and this type of test may be of more practical use in many conditions.

Demulsibility

Water inevitably gets into turbine lubricants, and into most other lubricants in circulatory systems, and the oil must separate readily from the water — that is, it must have a good demulsibility.

A standard test passes steam through a sample of the oil and measures the time the mixture takes to separate.

Additives purposely designed for stable emulsions are used in soluble cutting oils, in fire-resistant hydraulic fluids, and in some marine diesel cylinder lubricants.

Flash point

The flash point of an oil is the temperature at which it gives off a flammable vapour in specified conditions. Flash point has little significance for lubricant performance, but it is a useful check on possible contamination with a low flash product like petrol.

There are two main tests: for 'open' and for 'closed' flash point. The open flash point is higher than the closed one. Additives do not improve the flash point, and generally do not reduce it significantly.

Acidity and alkalinity

Acidity or alkalinity is now not very significant in new oils, particularly as additives such as ZDDP can considerably raise the acidity of new oil blends and detergent additives can significantly increase alkalinity. However, plotting the trend of acidity measurements is useful in following the degradation of some oils. If several results are plotted over a period in, say, a turbine oil, a sharp upward turn on the curve will be an indication of rapid oil oxidation. Acidity tests cannot be used simply in engine oils, where combustion products greatly affect acidity during the oil's life.

Acidity of lubricants is conventionally measured in terms of the amount of potassium hydroxide needed to neutralise the acidity – in mg of KOH/g of oil. Alkalinity is measured as the amount of KOH (again, as mg KOH/g of oil) that is equivalent to the acid needed to neutralise alkaline constituents in 1 gram of sample.

Alkalinity is deliberately introduced into some lubricants. Diesel engine oils may have alkaline additives to allow them to neutralise combustion acids from the fuel.

Other chemical and physical lubricating oil tests

This chapter has only touched on some of the great number of laboratory tests carried out on lubricating oils. There are full details of standard tests in the references cited.

One additional test can usefully be mentioned here: determination of electrical strength on transformer and switchgear oils. These are low-viscosity, stable oils, and they have to be quite exceptionally clean and free from moisture because they act as insulants under high voltage. Even a few parts per million of dirt or water can reduce the dielectric strength of a transformer oil below specification. A test for dielectric strength of oils is therefore a measure of condition more than of composition. The test is carried out by increasing a voltage across a 4 mm gap between two electrodes in the oil until a breakdown occurs. A British Standard for transformer oils gives details of the tests.

Detergent properties and laboratory engine tests

In the early days of diesel and petrol engines, plain mineral oils were sufficiently good lubricants. Later, oxidation inhibitors were used to increase oil life. Oil change periods were short — typically every thousand miles in cars — but the oils available were satisfactory over that period.

However, as specific power outputs increased, oxidation-inhibited oils became unsatisfactory as they did not prevent deposits in the engine that caused ring sticking and loss of power. Detergent properties were needed to keep combustion and oil-degradation products in suspension and prevent their depositing and baking in high-temperature zones. Low-temperature deposits also became a problem, particularly in petrol engines, and the ability of detergent-dispersant additives to deal with them was developed. Engine lubrication requirements are discussed in Chapter 3.

Figure 2.5 Engine testing is now highly automated and this row of control panels sets and monitors the operating conditions of a number of test engines

Types of engine tests

There are no conventional chemical or physical tests for detergent-dispersant performance; it is determined in standard test engines in closely controlled conditions. Widely used tests include those in the US Government MIL-L-2104D and MIL-L-46152 specifications.

Engines used include a special single-cylinder Caterpillar diesel in a number of variations. Special Petter diesel and petrol engines are used in Britain. Figures 2.5 and 2.6 show some typical engine-testing equipment. Oils are run in these engines for specified periods in controlled conditions and their quality is assessed by rating engine parts for freedom from deposits, resistance to ring sticking, and possibly bearing weight loss. In some tests, sludge formation on oil screens and rocker boxes and other characteristics will be measured.

Although this method is described as engine testing, what is being tested is, of course, the oil. The engine is only the means of doing this with more realism than laboratory tests allow, but with more controlled conditions and in a less time-consuming way than with production engines.

Figure 2.6   A Caterpillar test engine for lubricating oils. It is installed in a sound-proofed cell and connected to a dynamometer to absorb and measure power. It is controlled from the panels shown in figure 2.5

Methods of carrying out engine tests are described in detail in the Institute of Petroleum Methods for Analysis[6].

Results from these tests are internationally repeatable, and correlate well with service performance. However, when new oils are being developed, standard engine tests are always supplemented by controlled field tests in production engines, operated in normal working conditions. Results of engine tests are invaluable indicators of performance level, but not every engine test relates exactly to particular operating conditions.

The API (American Petroleum Institute) classification of engine oils[7] defines many engine test requirements, though it is more widely accepted in North America than in Europe (see Chapter 3).

European standards have been established by the Co-ordinating European Council, and are widely used[6]. Some tests now appear in international standards[8].

REFERENCES

1 *IP Methods for Analysis and Testing*, Heyden, London, 1982.
2 *ASTM Standards, Parts 17 and 18*, American Society for Testing and Materials, Philadelphia, Pennsylvania, 1982.
3 J. P. Allinson, *Criteria for Quality of Petroleum Products*, Applied Science Publishers (for Institute of Petroleum), London, 1973.
4 *SAE Handbook*, Society of Automotive Engineers, Warrendale, Pennsylvania, 1982.
5 *BS 4231: 1975 Viscosity classification for industrial liquid lubricants*, British Standards Institution, London.
6 *C.E.C. List of Methods*, Co-ordinating European Council, Institute of Petroleum, London, 1980.
7 *Engine Service Classifications and Guide to Crankcase Oil Selection, API Bulletin 1509*, American Petroleum Institute, Washington, D.C., 1980.
8 *ISO Catalogue*, section TC28, International Standards Organization, Geneva, 1982.

3 Diesel and Petrol Engine Lubrication

B. N. Seth *B.Sc., B.Sc. (Eng.), Ph.D., C.Eng., M.I.Mech.E.*
Esso Petroleum Company Limited

The lubrication requirements of high-speed diesel and petrol engines, and medium-speed trunk piston diesel engines, and the oil properties required for their satisfactory lubrication are reviewed in this chapter. The use of the more important official specifications for engine lubricants and the API Engine Service Classification System are discussed.

Diesel and petrol engines and their application can be classified as follows:

Type	Special range	Main application
Low-speed diesel	Below 300 rev/min	Main propulsion of large marine vessels by direct drive
Medium-speed diesel	300 to 1000 rev/min	Electric power generation, marine and railroad applications
High-speed diesel and petrol	Above 1000 rev/min	Commercial road vehicles, passenger cars, earth-moving and farm trade equipment

In this chapter only the medium-speed trunk piston diesel engines and the high-speed diesel/petrol engines are considered.

The lubrication of internal combustion engines is generally more difficult than that of most conventional machinery, owing to the wide variety of conditions — some of them extreme — encountered in their operation.

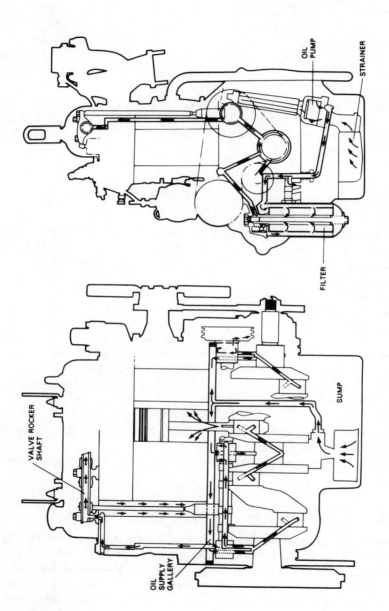

Figure 3.1 A typical lubricating system for a four-cylinder overhead valve engine

The lubrication problems in petrol engines are very much the same as those found in diesel engines, but are usually less severe and emphasise different aspects. The lubrication of the diesel engine power assembly (piston, piston rings and cylinder liner) is a more severe problem than the lubrication of its valve-actuating components. On the other hand, in a petrol engine the most severe problem is usually to be found in the lubrication of its valve train, particularly at the cam-to-cam follower interface. Petrol engines, particularly in stop/start motoring, are more prone to sludging.

The wear of cylinder liners, piston rings and grooves, and ring groove deposits are the major factors in engine maintenance. The trend towards higher output engines is well established: b.m.e.p. of 2 MPa (20 bar) is not uncommon in medium-speed diesel engines.

ENGINE LUBRICATION REQUIREMENTS

Engine lubrication systems are normally pressure-fed from an oil pump driven by the engine itself. However, some parts of the engine may be lubricated by gravity or splash from the action of the moving parts that dip into the sump oil and cause it to be flung to various internal engine parts. The pressure-fed system is used in the main to lubricate the bearings of the engine and valve-actuating components. The cylinder liners and piston assembly is normally lubricated by a generous supply of oil from splash. In some medium-speed trunk piston engines, provision is also made for additional lubrication of the cylinders by pressure-fed lubricant through the cylinder walls. Figure 3.1 gives a typical lubricating system for a four-cylinder overhead valve engine (without additional cylinder lubrication).

The lubrication requirements of internal combustion engines are considered under two headings:

(1) Combustion zone — includes cylinder liners, piston, piston rings and valve-actuating components.
(2) Crankcase — includes main crankshaft and connecting rod bearings.

Lubrication of combustion zone

Extremely high temperatures are experienced by the piston crown and upper cylinder walls, caused by the hot combustion gases. This heat from the piston crown is dissipated by conduction to the piston ring zone through the rings and thence through the cylinder walls, which are externally cooled. The oil film on these surfaces is therefore subjected to extreme heat and tends to decompose to form deposits. The products of combustion and oil degradation give rise to undesirable piston deposits and wear.

Piston deposits

Combustion products are rarely completely gaseous. Incompletely burnt fuel residues form gummy deposits which adhere to the combustion chamber walls. They combine or collect in the annular spaces between piston and cylinder walls where they come into contact with the lubricant. Fuel residues can be highly reactive and accelerate the deterioration of the oil in the hot conditions of the ring zone. If these oxidation products remain in the ring zone, deposits may build up on the ring and ring groove sides, increasing in thickness until the clearance between ring and groove is filled. Under these conditions the ring loses its freedom to move and ring sticking occurs. Similar deposits can arise from oil-degradation products.

Deterioration products may also build up in the space behind the ring. If the space becomes filled, the deposits push the ring radially outward resulting in high rates of ring and liner wear. Loss of ring movement is frequently the cause of ring leakage and 'blow-by'. In the larger diesels which burn residual fuel, incompletely burned fuel residues are heavy and gummy and may deposit in appreciable quantities.

Wear

There are many factors that contribute to the cylinder liner and piston ring wear. These vary to a greater or lesser degree with the design and operating conditions of the engine in question, but can be broadly classified as follows:

Mechanical wear — caused by metal-to-metal contact between moving parts under boundary lubrication conditions.
Corrosive wear — caused by the acidic products of combustion.
Abrasive wear — caused by hard foreign matter drawn into the cylinder with the air charge and by hard particles of carbon, asphalt, wear debris and ash-forming constituents from the fuel or lubricant.

All three causes are undoubtedly inter-related and can occur simultaneously, particularly in a diesel engine. Whilst these causes are accepted as normal wear hazards, their individual or multiple effects can be controlled by proper attention to operating conditions and selection of an appropriate lubricant.

In internal combustion engines, piston ring and cylinder liner surfaces and the valve-actuating components are important wear-prone areas. Diesel engines are generally more critical in regard to piston ring and liner wear, while petrol engines emphasise the problem of valve train wear.

Mechanical wear

Boundary lubrication conditions generally exist between piston ring and cylinder liner surfaces. This is particularly so at the top of the stroke where the piston momentarily comes to rest before reversing direction. The problem of maintaining an oil film at this point is further aggravated by the high temperatures around the upper cylinder. The combination of low sliding speed and higher temperature, combined with high gas pressures, effectively prevent a complete oil film being produced at the piston ring/cylinder liner interface. Some degree of metal-to-metal contact must therefore inevitably occur. As it occurs the surface asperities interlock, some damage being caused. The major asperities break off or melt and flow into the neighbouring depressions, thus causing mechanical wear.

Wear of valve-actuating components can be a serious problem, particularly in petrol engines. The cams on the camshaft are subject to large forces caused by the valve train inertia in modern high-speed petrol engines. Boundary lubrication conditions can exist resulting in metal-to-metal contact causing wear. With some combinations of malleable irons, the wear can take the form of tappet pitting, but with steel the cam noses in severe cases can be scuffed.

In general, under boundary lubrication conditions, the more viscous the lubricant the better it is. However, it is usually better to improve the lubricating properties of the oil by the addition of suitable additives into relatively low-viscosity base oils in order that the good cooling properities of the oil ar maintained.

Corrosive wear

It is generally agreed that corrosion is a major factor in cylinder liner wear. In diesel engines the corrosion is essentially due to the formation of sulphuric acid produced by the sulphur trioxide and water vapour within the engine cylinders as a result of the combustion process. The corrosive wear aspect is particularly important in medium-speed trunk piston diesel engines operating with high sulphur residual fuels (sulphur content of 3–4 per cent is not uncommon), for example, in electric power generation and marine applications. Fortunately the greater portion of sulphur oxides formed is sulphur dioxide, which is relatively non-corrosive.

In petrol engines the corrosion arises mainly from halogen oxy-acids produced as a result of the oxidation of the lead scavengers in petrol, during combustion.

At first glance it would appear that condensation in the presence of water vapour could not occur in the engine cylinders where gas temperatures can exceed 1670 °C during combustion. However, such condensations can and do occur in the cylinder. The cylinder liner skin temperature rarely exceeds 260 °C (in some cases it can be considerably lower) since the cylinder walls are cooled externally, usually by water. Sulphur trioxide even in small proportions raises the dew point (condensation temperature) of the combustion gases quite considerably. The combined effect of cylinder pressure and the presence of sulphur trioxide thus creates conditions where it is extremely likely that condensation of water vapour will occur. Sulphur trioxide has a strong affinity for water and it will readily combine with the condensed water to form corrosive sulphuric acid. The cylinder jacket temperatures may be directly related to the degree of corrosive wear experienced. In general, raising the temperature of the cooling fluid in the cylinder liner jacket can minimise the problem.

Abrasive wear

Abrasive wear is caused by hard particles of foreign matter being drawn into the engine with the air charge or via the crankcase breather system. When low-quality residual fuels are used, ash-forming constituents in such fuels will also materially contribute to increased abrasive wear.

A further factor contributing to abrasive wear is hard carbon formed by partial combustion of both fuel and lubricating oil. Incomplete combustion may result in carbon and asphaltenes, and inevitably some of these products will find their way to piston ring grooves. These deposits vary considerably in nature and quality according to the type of fuel and oil in use, but even soft carbonaceous deposits will become hard and abrasive when combined with sulphur compounds which can derive from the products of combustion.

If the oil film on the cylinder walls is attacked by sulphur oxides or acids, its lubricating properties will be impaired and the working surfaces will be made more vulnerable to the attack of abrasive materials.

Modern lubricants are designed to hold the combustion products, which could give rise to abrasive wear, in suspension, and they can thus be collected in the filtration system.

Crankcase lubrication

The lubrication environment is much less severe in the crankcase than in the combustion zone. In the crankcase the lubricant is subjected to high pressures in the main crankshaft and connecting rod bearings. The bulk oil temperature rarely exceeds 75 °C in the crankcase, so serious oil deterioration rarely results from the chemical changes in the oil itself. However, in some highly rated diesel engines, particularly those used in rail traction, the engines may be heavily loaded and inadequately cooled with the consequence that the crankcase oil temperatures may exceed 90 °C. Appreciable oil oxidation may then occur which produces deposits as varnish on piston walls and as gummy sludge on oil filters and oil lines.

Troubles may also arise when an engine is operated at relatively low temperatures, particularly when much of the operation is at no load or idling, a condition frequently encountered in automotive diesel and petrol engines under stop/start conditions. Combustion is usually poor during idling and contamination of the lubricating oil with unburned fuel is heavy.

These chemically unstable residues combine with water, condensed from blow-by gases passing the piston rings, to form gummy deposits which tend to separate in the form of varnish on rubbing surfaces and sticky sludge on oil filters, oil lines and other surfaces.

Cold running may allow considerable moisture to condense within the crankcase. This moisture becomes incorporated with other adventitious matter into the sludge, increasing its volume. Acidic moisture attacks rubbing surfaces causing increased wear during operation; it promotes rusting when the engine is shut down and the rust, being abrasive, could be responsible for added wear when the engine is restarted.

In extreme cases these gummy deposits can lead to oil line blockage causing lubricant starvation, leading to bearing failure.

The problem of corrosion of certain bearing materials should also be borne in mind. Here, certain lubricant formulations may be incompatible with particular bearing materials (for example, phosphor bronze, white metal) and can give rise to bearing corrosion.

LUBRICATING OIL PROPERTIES

The primary function of a lubricant is to lubricate; that is, it must minimise friction and wear between moving parts. In addition it must cool working parts and should have the ability to prevent or minimise corrosion and wear. It should possess good resistance to oxidation and thermal degradation. It should have good cold-starting properties — particularly important in automotive diesel and petrol engines.

Many of the necessary properties are inherent in the base oils from which the lubricating oils are manufactured. However, these in themselves are not sufficient to meet the lubrication requirements of modern highly rated engines. Chemicals commonly known as additives are added to the base oils to enhance the desirable natural properties, to modify undesirable properties or to impart completely new properties.

Thus a modern lubricating oil is a balanced formulation made up of carefully selected base oils which are normally highly refined petroleum-based mineral oils plus additive.

A modern engine lubricant should have good oxidation stability, resistance to corrosion, anti-wear and detergent-dispersant properties; it should have good low-temperature fluidity and anti-foamant properties. Additives that impart these properties broadly fall into three groups: those that keep the engine clean; those that reduce wear; and those that modify the physical properties of the oil. The type of additives used in engine lubricants and their function are summarised below:

(a) Detergent-dispersants, anti-oxidants — these are additives to reduce engine deposit and sludge.
(b) Basic detergents, anti-oxidants — these are additives to reduce corrosive wear.
(c) Anti-wear and EP agents — these are additives to reduce mechanical wear.
(d) VI improvers, pour depressants, anti-foamants — these are additives to modify base oil properties.

As discussed earlier, the lubricant in the combustion zone has to work in a very hostile environment giving rise to severe oxidative attacks. This is particularly true in the piston ring zone area where the first ring groove temperatures can easily reach 250 °C. This gives rise to acids, which promote corrosion of engine parts, and insoluble polymeric materials, which promote piston deposits. Anti-oxidants help to combat both these effects. Detergent-dispersants keep the insoluble products in suspension in the oil and thus prevent the formation of deposits in critical engine parts. The detergents with high basicity help in minimising corrosive wear. Anti-wear additives help in reducing mechanical wear of heavily loaded components, such as piston rings/cylinder liners and valve trains. Viscosity index improvers impart suitable viscosity characteristics, particularly important in automative petrol and diesel engines for cold starting. Other additives are added to improve low-temperature fluidity and foaming characteristics of the lubricating oils.

CLASSIFICATION OF ENGINE LUBRICANTS

Engine oil quality is defined by performance in full-scale or laboratory engines as well as in laboratory tests that are intended to predict performance under actual service conditions. There are two principal systems of engine oil classification:

(1) The SAE Viscosity Classification System classifies oil according to the viscosity at 100 °C and at various low temperatures depending on the viscosity grade. High-temperature viscosity is related to an oil's consumption and wear characteristics; low-temperature viscosity predicts the cold-starting performance and low-temperature lubrication. Oils with a high viscosity index are usually less sensitive to temperature changes and are therefore better able to perform efficiently at both high and low temperatures. Viscometric properties are also important for fuel economy.

(2) The API Engine Service Classification System rates engine oils in terms of their performance in selected engines operating under carefully controlled conditions designed to simulate severe service in the field. This system covers a wide range of service categories, including a number of prior as well as current military and industry engine tests.

Oils may also be classified according to specific military or industry specifications such as MIL-L-2104D, MIL-L-46152B or the now outdated Caterpillar Series 3. Details of the classification systems are discussed below.

SAE VISCOSITY CLASSIFICATION

The Society of Automotive Engineers (SAE) classifies engine oils by ten viscosity grades. The latest classification, SAE J300 SEP80, was approved in September

Table 3.1   SAE Viscosity Grades for Engine Oils, SAE J 300 SEP80[a]

SAE viscosity grade	Maximum viscosity (cP) at specified temperature[b]	Maximum border-line pumping temperature[c] (°C)	Viscosity (cSt) at 100 °C[d]	
			Minimum	Maximum
		(°C)		
0W	3250 at − 30 °C	−35	3.8	−
5W	3500 at − 25 °C	−30	3.8	−
10W	3500 at − 20 °C	−25	4.1	−
15W	3500 at − 15 °C	−20	5.6	−
20W	4500 at − 10 °C	−15	5.6	−
25W	6000 at − 5 °C	−10	9.3	−
20	−	−	5.6	< 9.3
30	−	−	9.3	<12.5
40	−	−	12.5	<16.3
50	−	−	16.3	<21.9

[a] Note that 1 cP = 1 mPa s; 1 cSt = 1 mm^2/s.
[b] Cold cranking simulator (ASTM D 2602).
[c] Mini-rotary viscometer (ASTM D 3829).
[d] Kinematic viscosity (ASTM D 445).

1980. After an eighteen-month optional-use period, it fully superseded the prior SAE J300d in March 1982. The limits for the fully metricated SAE J300 SEP80 are given in Table 3.1.

The W grades are based on a maximum low-temperature viscosity and a maximum border-line pumping temperature, as well as a minimum viscosity at 100 °C. Oils without the letter W are based on viscosity at 100 °C only. A multigraded oil is one whose low-temperature viscosity and border-line pumping temperature satisfy the requirements for one of the W grades, and whose 100 °C viscosity is within the prescribed range of a higher, non-W, grade. The low-temperature viscosity is determined by ASTM D 2602 (the tentative multi-temperature Cold Cranking Simulator (CCS) procedure is given in SAE J300 SEP80 until ASTM approves the revised D 2602). This CCS procedure determines the apparent dynamic viscosity of engine oils at low temperatures and rather high shear rates, and reports the results in centipoises. Viscosities measured by this method have been found to correlate with engine speeds developed during low-temperature cranking. Border-line pumping temperature is measured according to ASTM D 3829, using a mini-rotary viscometer at low shear rates, and the results are reported in degrees Celsius. The border-line pumping temperature is a measure of the ability of an oil to flow to the engine-oil pump inlet and provide adequate engine-oil pressure during the initial stages of operation.

Because engine cranking and starting, as well as oil flow, are important at low temperatures, the selection of an oil for winter operation should consider the viscosity required for successful cranking and starting, as well as the lowest ambient temperature expected. The oil viscosity at 100 °C is measured according to ASTM D 445, and the kinematic viscosity is reported in centistokes. Viscosities so measured are useful as a guide in selecting the proper viscosity oil for use under normal engine-operating temperatures.

Viscosity index

The viscosity of an oil changes with temperature. At low temperatures the oil is thick; its viscosity is high. As temperatures increase, the viscosity decreases. A sluggish oil makes engine starting difficult at colder temperatures and may not pump adequately to maintain a satisfactory oil pressure. On the other hand, oils of too low viscosity can give inadequate lubrication (causing wear) and high oil consumption.

Changes in oil viscosity with changing temperatures are not the same for all oils. One measure of viscosity change with temperature change is provided by the oil's viscosity index (VI) — the higher the VI, the less the viscosity change for a given temperature change. Addition of a VI improver will improve VI and its related viscosity-temperature performance characteristics (such as cold starting, fuel economy, and, in some cases, oil consumption and wear control), relative to those of single-grade oils with the same low shear viscosity at 100 °C.

Unfortunately, VI, which is based on kinematic viscosities at 40 °C and 100 °C, does not satisfactorily predict measured viscosities at low temperatures because of the combined effect of wax structure in the oil, VI improver, and other additives. Nevertheless, VI continues to be of some use for reflecting the performance of an oil as it relates to viscosity-temperature characteristics.

Multigrade oils

Although the viscosity-temperature quality of a single-viscosity grade oil is often sufficient to meet engine requirements for a particular season or climate, it may be inadequate in a different environment, necessitating a change to a grade better suited to the new conditions. In variable climate zones, motorists using single-graded oils generally change to a higher-numbered grade for summer driving and a lower-numbered grade for winter. Seasonal oil changes can be eliminated by the use of multigrade oils which, as discussed above, meet more than one grade in the SAE classification of crankcase oil viscosity.

Since each W grade (Table 3.1) is defined on the basis of a maximum viscosity and a maximum border-line pumping temperature, it is possible for an oil to satisfy the requirements of more than one W grade. A W-graded or a multigraded oil should be labelled with only the lowest W grade satisfied.

Thus an oil meeting the requirements for SAE grade 10W, 15W, 20W, 25W and 30 should be referred to as SAE 10W-30 grade only.

API SERVICE CLASSIFICATION SYSTEM (SAE J183)

In the early 1950s the American Petroleum Institute (API) introduced a system for classifying the various types of service conditions under which engines operate. These were designated as ML, MM, MS (for petrol engines) and DG, DM, DS (for diesel engines), but no performance standards were specified. In order to provide more precise definitions of oil performance and engine service, in 1969–70 API, in cooperation with the American Society for Testing and Materials (ASTM) and the Society of Automotive Engineers (SAE) established a new Engine Service Classification for engine oils. ASTM defined the test methods and performance targets. API developed the service letter designations and 'user' language. SAE combined the information into an SAE recommended practice in the SAE Handbook for consumer use. That report is called *Engine Oil Performance and Engine Oil Classification − SAE J183 FEB80*.

This current API Engine Service Classification is divided into an 'S' series covering engine oils generally sold in service stations for use in passenger cars and light trucks (mainly petrol engines), and a 'C' series for oils for use in commercial, farm, construction and off-highway vehicles (mainly diesel engines). An oil can meet more than one classification, for example, API SE, SF, CC.

The complete API system is described in API Bulletin 1509, eighth edition (revised 1980) *Engine Service Classification and Guide to Crankcase Oil Selection*. Table 3.2 gives a brief description of the API Engine Service Classification. Table 3.3 shows the relationships of the current classifications to the prior API service categories, and the corresponding military and/or major industry designations.

The API performance classification system is well known worldwide, and is widely used. In Europe, it is a requirement for service-fill engine oils, along with additional European targets.

CCMC REQUIREMENTS AND TEST FOR SERVICE-FILL ENGINE OILS

A group of European car makers has organised the Comité des Constructeurs d'Automobiles du Marché Commun (CCMC) to develop and adopt common engine and rig tests to define their performance requirements for service-fill engine oils for passenger cars and light trucks, and for diesel-powered commercial vehicles. Some of the engine performance requirements differ appreciably from the API SF, CC or CD targets. For some qualities ASTM engine sequence tests are used, either solely or as alternatives.

Table 3.2   API Engine Service Classification — summary

Letter designation	API Engine Service description	ASTM Engine Oil description
SA	Formerly for Utility Gasoline and Diesel Service	Oil without additive except that it may contain pour and/or foam depressants
SB	Minimum Duty Gasoline Engine Service	Provides some anti-oxidant and anti-scuff capabilities
SC	1964 Gasoline Engine Warranty Service	Oil meeting the 1964–67 requirements of the automobile manufacturers. Intended primarily for use in passenger cars. Provides low-temperature anti-sludge and anti-rust performance
SD	1968 Gasoline Engine Warranty Maintenance Service	Oil meeting the 1968–71 requirements of the automobile manufacturers. Intended primarily for use in passenger cars. Provides low-temperature anti-sludge and anti-rust performance
SE	1972 Gasoline Engine Warranty Maintenance Service	Oil meeting the 1972–79 requirements of the automobile manufacturers. Intended primarily for use in passenger cars. Provides high-temperature anti-oxidation, low-temperature anti-sludge and anti-rust performance
SF	1980 Gasoline Engine and Diesel Engine Service	Oil meeting the 1980 warranty requirements of the automobile manufacturers. Intended primarily for use in gasoline engine passenger cars. Provides protection against sludge, varnish rust, wear and high-temperature thickening

(continued overleaf)

Table 3.2   API Engine Service Classification — summary (continued)

Letter designation	API Engine Service description	ASTM Engine Oil description
CA for Diesel Engine Service	Light Duty Diesel Engine Service	Oil meeting the requirements of MIL-L-2104A. For use in gasoline and naturally aspirated diesel engines operated on low-sulphur fuel. The MIL-L-2104A specification was issued in 1954
CB for Diesel Engine Service	Moderate Duty Diesel Engine Service	Oil for use in gasoline and naturally aspirated diesel engines. Includes MIL-L-2104A oils where the diesel engine test was run using high-sulphur fuel
CC for Diesel Engine Service	Moderate Duty Diesel Engine Service	Oil meeting requirements of MIL-L-2104B. Provides low-temperature anti-sludge, anti-rust, and lightly supercharged diesel engine performance. The MIL-L-2104B specification was issued in 1964
CD for Diesel Engine Service	Severe Duty Diesel Engine Service	Oil meeting Caterpillar Tractor Co. certification requirements for Superior Lubricants (Series 3) for Caterpillar diesel engines. Provides moderately supercharged diesel engine performance. The certification of Series 3 oil was established by Caterpillar Tractor Co. in 1955. The related MIL-L-45199 specification was issued in 1958

Table 3.3 The relationships between API Engine Service Classifications and military and/or industry designations

Current API Engine Service Classifications	Previous API Engine Service Classifications	Related designations, military and industry
Service Station Engine Service		
SA	ML	Straight mineral oil ⎱ can have pour and foam depressants
SB	MM	Anti-wear-inhibited ⎰
SC	MS (1964–67)	1964 MS warranty approved: Ford M2C101-A; GM 4745-M
SD	MS (1968–71)	1968 MS warranty approved: Ford M2C101-B; GM 6041-M (before July 1970)
SE	None	1972 warranty approved: Ford M2C101-C, M2C153-A, and M2C157-A; GM 6136-M (SE) and 6146-M (SE/CC and SE/CD)
SF	None	1980 warranty approved: Ford M2C153-B; GM 6048-M (SF) and 6049-M (SF/CC and SF/CD)
Commercial and Fleet Engine Service		
CA	DG	MIL-L-2104A
CB	DM	Supplement 1
CC	DM	MIL-L-2104B (SC/CC), MIL-L-46152A (SE/CC), and MIL-L-46152B (SF/CC); GM 6146-M (SE/CC and SE/CD) and 6049-M (SF/CC and SF/CD)
CD	DS	MIL-L-45199B; Series 3; MIL-L-2104D; MIL-L-2104C (CD and SC); GM 6146-M (SE/CC and SE/CD) and 6049-M (SF/CC and SF/CD)

Table 3.4 lists current CCMC quality parameters, engine and test procedures for evaluating the service-fill engine oil requirements for passenger cars and diesel-powered commercial vehicles.

At the present time (1983), new CCMC proposals are under discussion, and it is likely that, in the near future, some considerable changes will be made to the requirements given in Table 3.4.

Table 3.4   CCMC Requirements and Tests for
Service-fill Engine Oils[a]

Performance parameter	Test	Test procedure
Passenger car service		
Low-temperature sludge	Fiat 600D	CEC L-04-A-70
	or	
	Sequence V-D	ASTM STP 315H
Bearing corrosion	Petter W-1	CEC L-02-A-78
	or	
	CRC L-38	Federal 791
High-temperature oxidation	Petter W-1	CEC L-02-A-78
	or	
	Sequence IIID	ASTM STP 315H
High-temperature deposits	Ford Cortina	CEC L-03-A-70
Pre-ignition	Fiat 124AC	CEC L-09-T-71
Wear	Daimler-Benz OM 616	CEC-L-17-A-78
Rust	Sequence IID	ASTM STP 315H
Shear stability[b]	Bosch injector	CEC L-14-A-78
	and	
	Peugeot 204	CEC L-25-A-78
Diesel-powered commercial vehicles		
High-temperature problems	MWB 'B'	CEC L-12-A-76
	and	
	Petter AVB	CEC L-24-A-78
Low-temperature sludge	Fiat 600D	CEC L-04-A-70
	or	
	Sequence V-D	ASTM STP 315H
Wear	Daimler-Benz OM 616	CEC L-17-A-78
Bearing corrosion	Petter W-1	CEC L-02-A-78
	or	
	CRC L-38	Federal 791
Rust	Sequence IID	ASTM STP 315H
	(being considered)	
Bore polishing	Volvo TD 120A	Volvo (150h)
	or	
	Ford Toronado	
	(being considered)	
Shear stability[b]	Bosch injector	CEC L-14-A-78

[a] Sources: *CCMC European Oil Sequence – For Service Fill Engine Oils for
          Passenger Cars*. 3rd Issue, Reference L/41/80, December 1980.
          *CCMC European Oil Sequence – For Service Fill Engine Oils for
          Diesel Powered Commercial Vehicles*. 5th Issue, Reference
          L/42/80, December 1980.

The revision of CCMC requirements is under discussion.

[b] Multigrade oils only.

HEAVY-DUTY OIL CLASSIFICATION

The simpler routine physical and chemical tests on lubricants yield only limited information on performance, and thus, in order to be able to predict the performance of a lubricating oil in service, it is necessary to resort to tests in actual engines. Over the years, various specifications for heavy-duty oils have been issued by both government agencies and engine builders. Requirements for these specifications reflect increasingly severe standards for engine cleanliness and other oil-performance qualities as determined by laboratory engine tests.

Better oils are called for as engine operation becomes more and more severe. Table 3.5 lists the commonly quoted official specifications and the engine tests involved in approximate order of increasing engine test severity and increasing oil detergency. Although some of these specifications are obsolete they are still useful in defining general quality level.

Engine test requirements

Engine tests are used to determine many characteristics of crankcase oils. Much of the cooperative research in developing test procedures has been done under the Co-ordinating Research Council (CRC), the American Society for Testing and Materials (ASTM), and the Co-ordinating European Council (CEC) in conjunction with government and other laboratories in both Europe and the USA. Standardised engine test procedures, including several originally developed by the Caterpillar Tractor Co., are now used by all major oil companies and are incorporated in most crankcase oil specifications. Brief descriptions of more widely used tests are given below.

Test for oil detergency

Caterpillar 1-A (CRC L-1)

Diesel, normally aspirated four-stroke single-cylinder test engine. Fuel sulphur content 0.4 per cent, 480 hour endurance test. Evaluation of wear, ring sticking and deposits. It can no longer be run.

Caterpillar 1-A Modified (CRC L-1 Modified)

Same as above, except fuel sulphur content is 1 per cent (0.95–1.05 per cent) used for 'Supplement 1' oils. Test also known as 'L-1 with 1 per cent Sulphur or Supplement 1 Test'. It can no longer be run.

Caterpillar 1-D

Similar to Caterpillar 1-A Modified (1 per cent sulphur), but with supercharged engine and at higher speed and temperature. 480 hour endurance test originally used to determine oil meeting Series 2 quality, but now used with 1-G for Series 3 oil.

Table 3.5   Official specifications and associated engine tests in approximate order of increasing test severity

Specification (in order of increasing oil detergency)	Engine tests required										
	Diesel (increasing severity →)						Petrol				
	L-1[a,b]	Mod. L-1[a,c]	Petter AV-1[c]	1-D[a,c]	1-H2[b]	1-G2[b]	L-38	Petter W-1	IID	IIID	V-D
MIL-L-2104A[d]	X	–	–	–	–	–	X	–	–	–	–
Supplement 1	–	X	–	–	–	–	X	–	–	–	–
DEF-2101-D[d]	–	–	X	–	–	–	–	X	–	–	–
DEF-STAN 91-43/1	–	–	X	–	–	–	–	X	–	–	–
MIL-L-2104B[d]	–	–	–	–	X	–	X[e]	–	X	–	–
MIL-L-46152[d]	–	–	–	–	X	–	X	–	X[f]	X[f]	X[f]
MIL-L-46152B	–	–	–	–	X	–	X	–	X	X	X
Series 2[d]	–	–	–	X	–	X	–	–	–	–	–
Series 3[d]	–	–	–	X	–	X	X	–	–	–	–
MIL-L-45199B[d]	–	–	–	X	–	X	X	–	–	–	–
MIL-L-2104C	–	–	–	–	–	X	X	–	X[g]	X[h]	X[g]
MIL-L-2104D (Note 1)	–	–	–	–	–	X	X	–	X[g]	X[h]	X[g]

[a] Obsolete engine test.
[b] 0.4 per cent sulphur fuel.
[c] 1.0 per cent sulphur fuel.
[d] Obsolete specification.
[e] Plus low-temperature deposit test.
[f] MIL-L-46152A issued January 1980 at the same basic quality level, but permitting re-refined as well as virgin base oils, and imposing a 0.14 mass per cent (max.) P limit plus wear limits in Sequence V-D.
[g] Less severe performance targets than for MIL-L-46152A and B.
[h] For valve train wear only.

Note 1.   In addition to the tests indicated in the table, MIL-L-2104D specification requires four new performance tests. These are: Detroit Diesel 6V-53T, Allison C-3 Friction Retention, Caterpillar TO-2 and Allison C-3 seal tests.

In 1980 this test was dropped by the Military, ASTM, SAE and API as no longer required for MIL-L-2104C or API CD because oils that passed Caterpillar 1-G/1-G2 tests also passed 1-D tests.

Caterpillar 1-G and 1-G2

Diesel supercharged, four-stroke single-cylinder engine test with 0.4 per cent sulphur fuel, 480 hours endurance test. Most severe of current MIL diesel tests. High temperatures and loads. Formerly used with 1-D for MIL-L-2104C oils, but since 1980 it is the only diesel test for MIL-L-2104C or API CD. 1-G and 1-G2 engines and test conditions are the same, except that 1-G2 uses a newer piston design, for availability reasons. Limits are different for the two tests. Since April 1983 this test is also a requirement for MIL-L-2104D.

Caterpillar 1-H and 1-H2

Uses 0.4 per cent sulphur fuel; same engine as 1-G; more severe conditions than L-1 but less severe than 1-G. A requirement of the obsolete MIL-L-2104B, 46152, 46152A and the current 46152B.

Caterpillar 1-H and 1-H2 engine and test conditions are the same, except that 1-H2 uses a newer piston design, for availability reasons. Limits are different for the two tests.

Detroit Diesel 6V-35T

Diesel supercharged two-cycle multi-cylinder engine with 0.4 per cent sulphur fuel, 240 hour endurance test. Evaluates diesel deposits, valve condition, engine scuffing and ring distress. This is included in the MIL-L-2104D specification, issued April 1983.

Petter AV-1 (Modified)

A single-cylinder, four-stroke diesel engine test, modified for greater severity as a requirement for the former DEF-2101-D and current DEF STAN 91-43/1 qualification. Modifications include 1 per cent sulphur fuel, no topping-up of the oil level during this test, and an oil-consumption limit of 0.030 lb/h (0.014 kg/h) for all grades.

Tests for oil oxidation and bearing corrosion

L-38 (CRC)

Uses a specially designed single-cylinder gasoline test engine (CLR Labeco Oil Test Engine). Evaluates oxidation and bearing corrosion characteristics of heavy-duty oils. Pertinent to all API Service Categories except SA.

Petter W-1

A single-cylinder gasoline-engine test, widely used in Europe. Evaluates oxidation and bearing-corrosion characteristics of heavy-duty oils. A requirement for current CCMC oils in Europe and former DEF-2101-D and current DEF STAN 91-43/1.

ASTM Sequence Engine tests

A brief description of ASTM Sequence Engine tests currently in use to define the API performance, and which are also included in the US Military Specifications as given below. It is to be noted that the test engines used are gasoline engines. *Sequence II-D.* Uses Oldsmobile V-8, 350 in.^{3} engine, and is primarily a rust test for MIL-L-46152B, MIL-L-2104C and API SE, SF quality oils.

Sequence III-D. Uses Oldsmobile V-8, 350 in.^{3} engine. The test is chiefly concerned with oil thickening (viscosity increase) under very hot engine conditions, and valve train wear for MIL-L-46152B, SE and SF oils. However, the performance targets differ for SE and SF levels. Sequence IIID is also run for MIL-L-2104C approval but for valve train wear only.

Sequence V-D. Uses Ford four-cylinder, 2.3 litre engine, and is primarily a test to evaluate low-temperature sludge and varnish control, plus in-take valve tip wear.

CCMC tests

For information, refer to Table 3.4.

Specification requirements

MIL-L-2104A (for API Service CA)

The now obsolete US MIL-L-2104A specification calls for engine tests on SAE 10W, 30 and 50 grade oils. It covers oil for both gasoline and diesel engines intended for general military use. Requirements include the relatively mild L-1 (0.4 per cent sulphur fuel) and L-38 tests.

Supplement 1 (for API Service CB)

This designation refers to an obsolete US specification that still has some use in the industry to describe oils of detergency level higher than MIL-L-2104A. Supplement 1 oils are tested similarly to MIL-L-2104A oils, except that the Modified L-1 test (1 per cent sulphur fuel) is used instead of the L-1.

DEF-2102-D (no API category but falls within API CB)

The obsolete British DEF-2102-D specifies a Petter W-1 gasoline engine test and a modified Petter AV-1 diesel engine test using 1 per cent sulphur fuel. The Petter W-1 assesses oil oxidation-resistance and bearing-corrosion properties. The Petter AV-1 assesses engine cleanliness (high temperature) and wear.

DEF STAN 91-43/1 (no API category but falls within API CB)

This British specification replaced DEF-2102-D and uses the same Petter W-1 and AV-1 engine tests. It covers four viscosity grades for certain compression-ignition and gasoline engines normally satisfied by SB/CB quality oils.

MIL-L-2104B (for API Service CC)

The obsolete US military specification MIL-L-2104B is a more stringent version of MIL-L-2104A. Performance tests include the more severe Caterpillar 1-H or 1-H2 diesel engine tests (0.4 per cent sulphur fuel) and the L-38, together with a low-temperature deposit test and a rusting test in Labeco and Oldsmobile engines, respectively.

MIL-L-46152 and 46152A (for API Service SE and CC)

MIL-L-46152, issued in November 1970, covers engine oils for civilian government vehicles, including passenger cars, light-duty to medium-duty trucks, and lightly supercharged diesel engines in moderate duty. It favours US carmakers' warranty performance targets for gasoline engine oils but provides for Caterpillar 1-H or 1-H2 (MIL-L-2104B) level of diesel engine performance. These oils meet API SE and CC categories.

MIL-L-46152A was issued in January 1980 and differs chiefly in two respects:

- For the first time, it permitted re-refined oils as well as virgin base stocks.
- It substituted Sequence IIID for IIIC with equivalent performance targets.

MIL-L-46152B (for API Service SF and CC)

MIL-L-46152B was issued in January 1981. It represents an upgrading of MIL-L-46152A to API SF and CC quality to provide warranty-quality oil for newer gasoline engines, as well as to continue to service the same types of diesel engines. Virgin or re-refined base stocks can be used.

Series 2 (for API Service CD)

This Caterpillar specification described heavy-duty oils intended for the most severe diesel engine conditions. Series 3 oils had to pass the Caterpillar 1-D diesel engine test with 1 per cent sulphur fuel, plus the more highly supercharged 1-G test. Caterpillar discontinued its Series 3 approvals in October 1972 and now simply recommends oils of the API CD level, which is satisfied by the obsolete Series 3, obsolete MIL-L-45199B, and current MIL-L-2104C qualities.

MIL-L-45199B (for API Service CD)

The engine test requirements for this US Government specification are the same as those for Series 3 oils, except that the oils must also pass the L-38 gasoline engine test. Series 3 and MIL-L-45199B represent the same level of diesel detergency. This specification was superseded by MIL-L-2104C.

MIL-L-2104C (for API Service SC and CD)

This US Military specification, issued in November 1970, covers engine oils for all types of reciprocating internal combustion engines of both spark-ignition and compression-ignition types in tactical service. It favours performance in turbocharged diesel engines and older MIL-L-2104B (API SC) level of deposit control in gasoline engines. This specification has been superseded by MIL-L-2104D.

MIL-L-2104D

This US Military specification, issued in April 1983, covers engine oils suitable for lubrication of reciprocating internal combustion engines of both spark-ignition and compression-ignition types, and for power transmission fluid applications in equipment used in tactical service, and for the first time allows a multi-viscosity oil (SAE 15W-40).

MIL-L-9000G

This specification covers a single-grade lubricating oil, in the SAE 30-40 range suitable for use in advanced design, high out-put, shipboard main propulsion and auxiliary diesel engines. The engine tests required are GM-71 and Caterpillar 1-G2 Modified (1.0 per cent sulphur fuel, no oil drain).

DEF STAN 91-22/2 (OMD 113) (no API category but falls within API CD)

This UK Ministry of Defence standard relates to lubricating oils intended for use in land and marine diesel engines, rangine from small single-cylinder naturally aspirated type to multi-cylinder highly rated pressure-charged engines. It calls for a single-grade oil in the SAE 30–40 range. The engine acceptance tests are Petter W-1, Petter AV-B and Rootes TS-3 two-stroke engines.

SELECTION OF ENGINE LUBRICANTS

Engine design and construction, fuel, operating conditions and maintenance practices are important factors for consideration in the selection of engine lubricants for a particular application.

The various official specifications (Table 3.5) and the new API Engine Oil Performance Classification System (Tables 3.2 and 3.3) discussed earlier can be used as a guide in the selection of engine lubricants.

Most engine manufacturers have their own engine oil specification(s) and approval system, and issue a list of recommended lubricants. In many cases these are related to the official specifications and/or API Engine Service Classifications. Again, these can be used as a guide to proper selection of engine lubricants.

The engine manufacturers' specifications, particularly high-speed diesel and gasoline engines, usually relate to minimum-quality requirements, and in general no harm will occur by the use of higher-quality oil.

Medium-speed diesel engines used in railroad, power generation and/or marine applications often have specialised requirements. Here the engine manufacturers' recommendations should be strictly adhered to.

The selection of a proper lubricant is a specialised job, and in all cases it is good practice to consult a major oil company for expert advice.

4 Lubrication of Water and Steam Turbines

M. G. Hayler *Ph.D., B.Sc., C.Eng., M.I.Mech.E.*
Esso Petroleum Company Limited
A. C. M. Wilson *A.M.C.T., F.Inst.Pet.*

The steam turbine is used as the prime mover for driving equipment such as blowers, pumps and compressors, but its predominant use is to drive alternators and generators for the production of electricity.

The size of turbines for this application has increased tremendously in the last three decades and while a 30 MW unit was the standard size built immediately post-war, 660 MW, and in some cases up to 1300 MW, tandem-compound machines are currently being constructed or installed in most advanced countries.

Water turbines are restricted in size and numbers by the availability of suitable water sources. They are used for electricity generation but are often associated with pump storage and irrigation schemes. The largest size in the UK is 300 MW.

The increase in the size of machines has resulted in higher bearing loads and journal peripheral speeds, and in order to reduce the size and increase the sensitivity of the control gear the hydraulic fluid pressures have been increased. The lubricant has therefore to perform satisfactorily under arduous conditions and has to withstand continuous thermal cycling and contamination by solid particles, sludge, water, air, hydrogen and process materials. It has to act as a bearing lubricant, as a control gear hydraulic fluid and, in steam turbines, predominantly as a coolant.

Although there are no arduous lubrication requirements such as require a load-carrying or anti-wear additive, turbine oils must be amongst the most technologically advanced lubricants to combat the conditions encountered.

TYPES OF TURBINE

Steam turbines

Steam turbines can be classified generally as: (a) large units used for driving alternators for the production of electricity[1]; (b) smaller units for auxiliary and industrial electricity generation; and (c) small industrial units used for driving equipment such as blowers, pumps and compressors.

Large units run at speeds dependent on the frequency of the electrical system, for example, 1500 or 3000 rev/min for a 50 Hz system and 1800 or 3600 rev/min for a 60 Hz system. Up to approximately 20 MW output single-cylinder (that is, one turbine stage) machines with two bearings are used, but at 500 MW five-cylinder machines with a larger number of bearings are normal and rigid couplings allow for one instead of two bearings between some cylinders.

Steam temperatures, which play an important part in the performance of the lubricant, range from approximately 200 °C to 600 °C. The second class of turbine ranges up to approximately 30 MW with speeds up to 10 000 rev/min for low-output machines using reduction gearing to the driven unit. The third class is similar to the small units mentioned above, but includes variable speed units.

Water turbines

The two main types of water turbine are the impulse and reaction (or pressure) turbines. The former, commonly known as Pelton wheels, are normally constructed with a horizontal shaft, and are operated by high-velocity water jets that impinge on buckets carried on the periphery of the turbine rotor. The latter are the most common type and are normally constructed with a vertical shaft. They have either an inward flow, or Francis type, runner or a propeller runner of the fixed or movable (Kaplan) blade type. Speeds range between approximately 70 and 1000 rev/min.

AREAS TO BE LUBRICATED

A turbine system consists essentially of three parts, each of which requires lubrication; these are the turbine, the turbine control gear and the driven unit.

Turbine lubrication

The main bearings of a steam turbine are plain white metal journal bearings[2] and are lubricated hydrodynamically by the formation of a high-pressure wedge of oil between the shaft and white metal. Because of the load on the bearings of large turbines at start-up, hydrostatic pressure lubrication is used to lift the shaft from the bottom of the bearing. This facilitates the formation of a hydrodynamic film and prevents wear.

The thrust bearings are usually of the multi-collar type on small turbines, but of the Michell tilting pad type on large machines. During running the pads tilt to allow the formation of an oil wedge and hydrodynamic lubrication. When the output of the turbine is transmitted through reduction gears, these are double helical and are spray or bath lubricated.

In all except the largest of machines, which are direct coupled, flexible couplings are used to join individual sections of shaft, and these require continuous lubrication to prevent severe wear. This is achieved by feeding oil from the end of the bearing for couplings that are adjacent to a bearing or by independent feed lubrication where the coupling is separated from the bearing housing.

A typical vertical water turbine has one runner, two alternator guides and one thrust bearing. The thrust bearing, which supports the weight of the machine, is the most critical to lubricate.

Control system lubrication

The control system of a steam turbine may be mechanical, hydraulic or electro-hydraulic. Except in the smallest units, which are wholly mechanically operated, the governor response is transmitted to relay or servo valves by oil pressure. Protective devices such as trip and unloading gear are also hydraulically operated.

Hydraulic fluid pressures range from $< 7 \times 10^5$ N/m^2 to $> 70 \times 10^5$ N/m^2, and although mineral oils are employed at low pressures it is customary to use a fire-resistant fluid at pressures above approximately 20×10^5 N/m^2. This is because of the proximity of steam pipes that are at temperatures higher than the auto-ignition temperature of mineral oil.

The control system of a water turbine is relatively simple. A mechanical governor, driven from the turbine shaft, operates a hydraulic valve which in turn operates a relay valve controlling the flow of oil to the gate-operating cylinders. The gate controls the flow of water to the runner, and, therefore, the speed of the turbine.

Driven unit lubrication

The lubrication requirements of the driven unit are often specialised, as in the case of compressors, and are dealt with elsewhere in this book. Driven units such as alternators and generators have similar requirements to the turbine itself, with journal and thrust bearings and oil seals common to most requirements.

In alternators cooled by hydrogen, the oil-sealed glands, which prevent hydrogen leakage, are lubricated by controlled oil leakage across the seal. Hydrogen contamination of the seal oil occurs and this is removed by natural or vacuum degassing.

LUBRICATION SYSTEMS

Water and steam turbines are almost exclusively lubricated by petroleum mineral oils and the systems described are not necessarily suitable for other types of lubricant.

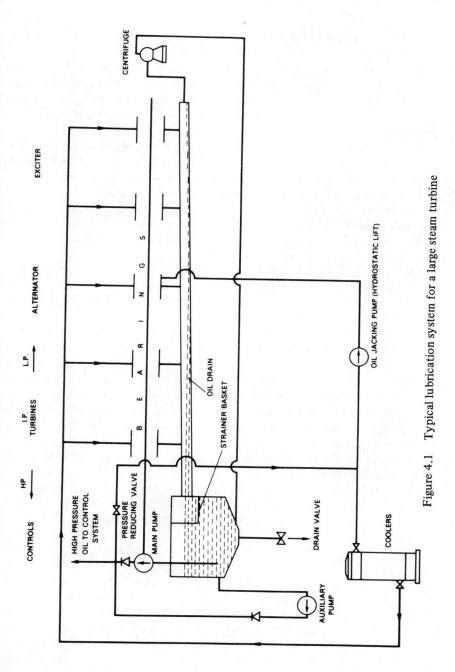

Figure 4.1 Typical lubrication system for a large steam turbine

A typical circulatory system for a large steam turbine is shown diagram-matically in figure 4.1. Essentially it consists of a reservoir, a pump, pressure-control devices, a cooler, and oil-purifying equipment.

One or two pumps, usually of the gear or centrifugal type and driven from the main shaft, are used to deliver oil at low pressure (between 0.3×10^5 and 1.5×10^5 N/m^2) and temperature ($40-50$ °C) to the journal and thrust bearings. Oil at higher pressure (between 4×10^5 and 20×10^5 N/m^2) is delivered to the governor and control system.

The system is designed for between five and ten passes of oil per hour and may contain from approximately 450 litres for a small machine to over 100 000 litres for the larger machines. Oil from the system returns to the tank at a tem-perature between 50 and 75 °C, by way of the pedestal drains and return pipe which are only part full. A strainer basket is situated in the tank at the end of the return pipe which enters horizontally at oil-surface level.

The tank has baffles to aid the separation of air and water and a sloping bottom so that water can be easily drained. The pump suction, often fitted with an ejector for priming centrifugal pumps, is situated higher than the drainage point. Oil for the purifier, which operates as a bypass system, is taken from one end of the tank and returned, often by way of the main oil return pipe, just below the oil surface at the opposite end of the tank.

Fine filters are fitted in the oil-supply lines to protect bearings and relay valves from wear and blockage by abrasive particles. A cooler is situated between the pump delivery and the bearings.

The tank is ventilated to remove water and acid vapour and this is achieved by natural or forced ventilation, or by the use of a cooler dehumidifier.

Steam glands and oil seals are an important part of the lubrication system as these have to prevent oil leakage and contamination of the oil by water (steam). Carbon and labyrinth seals are used to prevent steam leakage, and oil leakage is prevented by oil guards and thrower rings. The oil may be subjected to very high temperatures in these areas.

Small turbines have much simpler lubrication systems. Ring oiled bearings, gears and anti-friction thrust bearings are commonly used with oil bath lubrication. In order to remove heat from the oil the bath is double walled to allow water circulation.

Four basic systems are used for lubricating water turbines: gravity, unit, self-oiling, and oil bath; combinations of these are also used. The gravity system utilises a header tank to feed oil to the bearings; the oil then returns to a lower tank where it is filtered and then pumped back to the header tank. The unit system is a simplification of that described for a steam turbine, and the oil is pumped to the bearings from a main tank to which it returns. The self-oiling system is used on the lower alternator guide and utilises a shaft-driven pump to circulate oil from a sump which is integral with the bearing housing. The top guide and thrust bearing are then oil-bath lubricated. The runner bearing of all

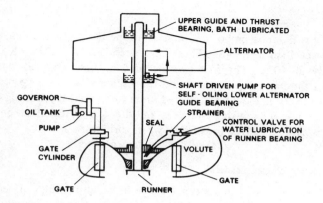

Figure 4.2 Typical mixed lubrication system for a vertical water turbine

vertical machines is usually lubricated independently by a self-oiling system, by grease or by filtered water.

The control system is supplied with high-pressure oil from a separate tank in the case of self-oiling systems, and from the drain tank in the case of gravity and unit systems. Figure 4.2 shows a typical mixed system for a vertical turbine.

LUBRICANT PROPERTIES

The petroleum mineral oils most commonly used in turbines contain oxidation and rust inhibitors: They nominally comply with the British Standard Specification BS 489 'Steam Turbine Oils' and their important properties, some of which are listed in Table 4.1, are described below.

Table 4.1 Comparison of significant properties of mineral oil and fire-resistant fluid

Property	Mineral oil	Fire-resistant fluid
Auto-ignition temperature (°C)	390	640
Viscosity (cSt) at 0 °C	460	1300
40 °C	46	46
100 °C	6	4
Density at 20 °C	0.870	1.200
Rust-preventing characteristics:		
Salt water	Pass	Fail
Distilled water	Pass	Pass

Viscosity

Viscosity[2] is the most important physical property of a lubricating oil and four grades are specified in BS 489. These have viscosities of 32, 46, 68 and 100 cSt respectively at 40 °C. The first two are used in direct-coupled machines, the second and third in water turbines and the third and fourth in geared machines. A higher viscosity is required in geared units to combat the loading on the gear teeth and so prevent scuffing and wear, and the load on the thrust bearing often decides the viscosity choice for vertical water turbines.

Rust-preventing characteristics

Three phases are present in a steam turbine in which rusting can occur if water or steam enters the system from a cooler leak or as a result of faulty operation or seals. These are the full oil flow, static water and vapour space phases.

Oils are formulated to pass the IP 135 'Rust preventing characteristics of steam turbine oil' test in the presence of a synthetic sea water. This assesses the oil in full flow and assures protection in service. Only a limited protection is given against corrosion under static water-film conditions and no protection is given to surfaces that are not covered or continually splashed with oil. So if moisture condensation occurs, or water films settle and lie on ferrous metal for any period of time, rusting will follow.

The effects of rusting can be very serious and, so far as is possible, water should be excluded from the system. Suitable surface coatings and non-corrosive materials are used to reduce corrosion in vapour spaces and sensitive areas such as control gear.

Water ingress to water turbines is by way of the runner seal and is negligible in water-lubricated runners where the seal is above the bearing. In the case of oil- and grease-lubricated runners the seal is below the bearing, and water passing the seal has to be vented away before reaching the bearing. Oils containing anti-rust additives are used, however, to combat the effects of atmospheric water condensation in sensitive areas.

Oxidation stability

Petroleum oils, which are a mixture of hydrocarbons, oxidise when subjected to high temperature and oxygen (air) with the formation of acids and a solid material that is familiarly known as sludge[3]. These corrode the system, block the valves and oilways and result in malfunctioning of the equipment.

Many materials act as oxidation catalysts (that is, they increase the rate of oxidation) and in a turbine system copper, a very active catalyst, is picked up by mechanical and erosion mechanisms from bearings, piping, coolers, etc. Certain iron compounds, which may be formed by the action of oil oxidation products and water on steel, are also very active catalysts.

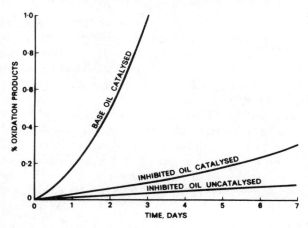

Figure 4.3   Oxidation curves for typical oils, using IP 280/80 test equipment

The oil is refined to a stage at which the addition of an oxidation inhibitor will further improve its stability. This anti-oxidant has also to counteract any adverse effects of the anti-rust additive. In addition to a conventional anti-oxidant, modern oils usually contain an additive that reduces the catalytic effect of metals, particularly those in solution.

The stability of turbine oil is assessed by a number of tests, the most important of which are IP 280 'Oxidation stability of inhibited mineral turbine oils' and an extended version of IP 306. These tests use oxygen/solid copper and oil-soluble copper and iron salts as catalysts and are completed in 164 hours at 120 °C. Typical oxidation curves are shown in figure 4.3. IP 157 is also important as a long-term turbine oil oxidation test.

The level of stability required in an oil is a question of individual technical and economic requirements. The oil must be sufficiently stable to withstand the conditions encountered during passage through the system and additionally to remain in acceptable condition for an economic period. The oil in small turbines may be changed after a period of one year or so, but in large turbines a life of up to thirty years is desirable because of the outage and oil change costs involved.

Although it is customary to use the same quality oil in water turbines as is used in steam turbines, stability is of less importance because of the low temperatures and small oil volumes.

Foaming and air retention

Foam and air retained in oil[4] lead to pump cavitation, loss of control and bearing damage. It is essential therefore that air is quickly removed from oil, and the system design precludes the formation of foam and air–oil emulsions. High-pressure oil passing through the valves, couplings, low suction heads and badly designed return lines are some of the causes of foam and entrained air.

The bubble size is the essential criterion that determines whether air creates foam or is entrained. Viscosity, and therefore temperature, and the rate at which air is injected into the oil are also important. The oil formulation can affect both properties and control is necessary. Anti-foam additives suppress foam but in doing so some types may prevent agglomeration of air bubbles and so slow down air release.

Tests designed to show the tendency of an oil to foam and retain air are the IP 146 'Foaming characteristics of lubricating oils' and the IP 313 method 'Determination of air release properties'. In the former air is bubbled slowly at low pressure into the oil and the amount and rate of dissipation of the foam measured, while in the latter air is injected into oil at high pressure and the amount of air entrained calculated from the change in oil density.

Demulsification

Water must readily separate from oil in the drain tank so that it is dry when pumped to the system. In a new oil this tendency is controlled by IP 19 'Demulsification number of lubricating oil' which measures the ability of water to separate from an emulsion formed by blowing steam into oil. Results on used oils are of less significance because of the effects of oxidation products and contamination. In service, oil treatment enables oils of relatively high demulsification numbers to be kept dry and perform satisfactorily.

Other properties

The corrosive effect of oil on copper and its alloys, and the water solubility of additives, are two further relatively important properties. However, both are

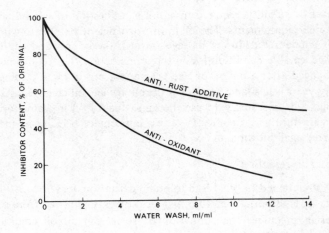

Figure 4.4   Effect of water washing on oil inhibitors

generally covered by other tests. Copper corrosion is specifically covered in BS 489, but is also covered indirectly by the more severe conditions of oxidation tests that use solid copper as a catalyst. Additive solubility in water is covered to a reasonable extent by the demulsification number and rusting characteristics tests, the results of which would be affected by water-soluble additives. Modern oils will withstand a high degree of water washing without impairment of the properties for which additives are included (figure 4.4).

LUBRICANT MAINTENANCE

The initial cleaning of the lubrication system and its maintenance in a satisfactory condition, together with that of the oil, are the most important factors leading to satisfactory turbine lubrication. The four most important oil properties to be maintained are cleanliness, dryness, rusting characteristics and stability.

In equipment operating on a high-pressure control system, clearances in valves and pumps are small and may be < 5 μm. Furthermore, the amount of erosion damage to pumps, valves and bearings is proportional to the particle size of detritus and its content of the oil. Scoring and failures of bearings[5] occur as a result of the presence of large detritus particles such as are present in a newly erected system. Normal system cleanliness, however, is such that the inorganic insolubles (mainly iron oxides) content of oil is < 0.002 per cent for a particle size > 5 μm nominal.

The rate of removal of water by drainage, centrifuge and vapour extraction must always exceed its ingress otherwise wet oil will be returned to the system and result in severe corrosion and malfunctioning of the equipment. A sustained high water content can also lead to bacterial and fungal growth in the system[6]. This can cause filter blocking and formation of deposits. Hardening and failure of white metal bearings[7] is accelerated by the presence of water and electrolytes in the oil. To control the corrosive effects of water in the fluid flow phase, and lessen its effects in the static water phase, an adequate degree of rust-preventing characteristics must be maintained. The rusting characteristics test is often used for this purpose and the oil re-inhibited when it fails to pass the test in the presence of distilled water.

The oxidation stability of oil decreases in service until eventually the acid value of the oil increases and sludge is formed. The initial change in stability is rapid but then slows down and proceeds as shown in figure 4.5. The determination of acid value is adequate for small oil volume systems where a quick oil change can be made at short notice, but for large volume systems an oxidation test can be used to give early warning of oil deterioration and the necessity for an oil change.

Because of the importance of oil quality and its effects on system performance in larger units, regular monitoring of oil samples is normal. Such monitoring

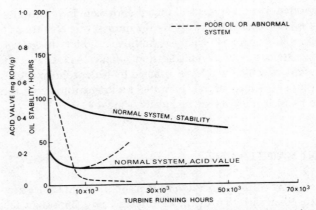

Figure 4.5   Acid value and oil stability changes in service

would, typically, check viscosity, water content, rusting characteristics and oxidation stability. It allows a gradual change in oil condition to be followed over a period. Any changes due to system malfunction can thus be detected at an early stage, allowing corrective action to be taken before a serious condition occurs.

SPECIAL-REQUIREMENT LUBRICANTS

Fire-resistant fluids

Fire-resistant fluids used in turbine control systems[8,9] are based on organic phosphate esters, the most common being tri-cresyl and tri-xylenyl phosphates. The control system is separate from the lubrication system and mineral oil is still used to lubricate the bearings. The cost of such fluids is approximately three times that of mineral oil.

In general the same properties are of importance as with mineral oils. Additionally these fluids undergo hydrolysis and form corrosive compounds in the presence of water. In consequence they have to be kept particularly dry and this is achieved by filtering through an absorbent earth filter. Their oxidation and thermal stabilities are not as good as those of mineral turbine oil and they must not be subjected to high spot temperatures, such as exist in control gear, for any significant period of time. Their viscosity/temperature characteristics are less satisfactory than those of mineral oil and heating may be necessary on cold start-up to ensure that the fluid can be pumped. As they also have a higher density the design of the pump and pipework has to be more liberal. They are particularly good solvents and surface-coating and seal materials have to be carefully chosen. Comparative figures for mineral oil and phosphate ester are given in Table 4.1.

Many properties of these fluids are superior to those of mineral oil, and in particular their bulk modulus, low compressibility and low vapour pressure ensure rapid control system response.

Pedestal feet lubrication

When a steam turbine heats up or cools down expansion and contraction occurs. The turbine moves and is carried by sliding pedestal feet, and in the event of the feet sticking the turbine shaft will bend. Lubrication of these feet in large turbines is difficult because of the loads and temperatures involved. Molybdenum disulphide incorporated in a lithium soap base grease is generally used for this purpose and is force-fed to the loaded areas.

REFERENCES

1   *Modern Power Station Practice, Volume 3*, C.E.G.B., London, 1963.
2   P. Ramsden, Lubrication and wear, fundamentals and application to design, *Proc. Inst. Mech. Eng.*, 182, Part 3A (1967) 75-81.
3   A. C. M. Wilson, Oxidation stability of steam turbine oils, *J. Inst. Petrol.*, 50, No. 482 (1964) 47-57.
4   P. D. Claxton, Aeration of petroleum based turbine oils, service and laboratory experience, *Tribology*, (February 1972) 8-13.
5   P. H. Dawson, F. Fidler and P. Rowley, *Lubrication and wear — third convention, Institute of Mechanical Engineers, London, 1965*, pp.27-36.
6   D. Summers-Smith, Microbial activity in industrial turbine lubrication systems, *Tribology International*, (August 1982) 180-186.
7   J. B. Bryce and T. G. Roehner, Corrosion of tin base Babbitt bearings in marine steam turbines, *Trans. Inst. Mar. Eng.*, 73, No. 11 (1961) 1-35.
8   A. C. M. Wilson, Fire resistant fluids for general hydraulic and steam turbine systems, *Proc. Inst. Mech. Eng.*, 182, Part 1, No. 5 (1967-68) 109-134.
9   G. F. Wolfe and M. Cohen, *Fire Resistant Fluids for Steam Turbine Electro-hydraulic Control Applications, STP No. 437*, American Society for Testing and Materials, Philadelphia, Pennsylvania, 1967.

5 Industrial Gas-turbine Lubrication

M. G. Hayler *Ph.D., B.Sc., C.Eng., M.I.Mech.E.*
Esso Petroleum Company Limited
A. T. Langton *C.Eng., M.I.Mech.E.*
L. F. Rutishauser *B.Sc., D.I.C.*

The adoption of the gas turbine for industrial applications became important in the early 1960s and the gas turbine is now firmly established as a prime mover for industry. Early installations used small gas turbines of up to 370 kW, mainly for military emergency generation. Emphasis on this type of unit has decreased and the market requirement has shifted towards larger units, some as large as 80 MW. Compact self-contained electrical power units remain a major application but medium-sized gas turbines are used in applications such as offshore oil rigs. Their compactness favours use there for driving compressors and pumps, and for electrical power.

CHARACTERISTICS

Gas turbines offer a number of advantages over other forms of prime movers: these include low capital cost, reliability, low maintenance requirements, compactness, quick start-up and ease of control. However, they have, until recently, tended to be at a disadvantage in terms of thermal efficiency compared with the pressure-charged diesel engine. Technical developments have narrowed the thermal efficiency gap, but the gas turbine is expected to remain generally more demanding on fuel quality.

Based on design and construction, gas turbines may be divided into two general categories: heavy industrial units and modified aero engine units.

Heavy industrial units

These are integral units that combine a compressor, a combustion system and a power output turbine in a single rugged casing. The unit is characterised by a comparatively low pressure ratio (about 10:1) and a large combustion chamber

Figure 5.1   Cutaway view of the Ruston Tornado gas turbine
(courtesy of Ruston Gas Turbines Ltd)

(or chambers) with relatively low speeds, cycle temperature and turbine blade stresses. Figure 5.1 illustrates a Ruston Tornado heavy-duty industrial gas turbine that develops 8500 horsepower.

Heavy industrial units can be sub-divided into two further types:

Single shaft version, in which the compressor, turbine and power take-off is on a common shaft. This arrangement is more commonly used for electrical generation where speed variations must be minimised.

Two shaft version, in which one turbine stage drives the compressor and a second stage independently provides the power take-off. This arrangement gives greater flexibility in load and speed and is more commonly used in the compressor (or marine) application.

Modified aero engine type

These units utilise one or more aircraft-type gasifiers driving a single, separate power output turbine. A typical arrangement is illustrated in figure 5.2, which shows a gas turbine plant used for gas compression at St Fergus, Scotland. The plant comprises eight gas turbine generator sets.

Figure 5.2   One of eight GEC gas turbine gas compressor sets at the British Gas North Sea terminal at St Fergus, Scotland; total power 105 MW (courtesy of GEC Gas Turbines Ltd)

Application

Which type is favoured depends on the significance of: capital costs, fuel consumption, turbine life, time taken to reach full load, and physical proportion of the machinery. The heavy industrial type is of relatively massive construction, with lower power/weight ratio, moderate operating conditions and low maintenance. In contrast, the aviation turbine has a higher power/weight ratio and operates under more severe conditions. Each type has its own unique features, and the industrial operator has considerable flexibility in choosing the gas turbine best suited to his particular requirements.

PERFORMANCE TRENDS

When gas turbines first entered industrial service, their inlet temperatures rarely exceeded 700 °C, with corresponding overall efficiencies of less than 13 per cent, which compared unfavourably with the industrial diesel engine. Gas turbines could however be made more economically attractive by operating on residual fuels. Later developments allowed turbine inlet temperature to be raised to 850–950 °C with a corresponding overall efficiency of 22–28 per cent. Even so, this represented the upper limit of operating temperature with uncooled turbine blades until 1970, when the introduction of cooled blades enabled inlet temperatures to be raised to 1250 °C. At this level, overall efficiency approaches 40 per cent, making the gas turbine fully competitive with the diesel. Given this parity, emphasis on the gas turbine's light weight, low space, and rapidity and ease of maintenance represents additionally attractive features. On the other hand, blade cooling technology is difficult to apply to the heavy industrial gas turbine, because of the large size of these units, and the mass of blade metal to be cooled. Thus heavy industrial gas turbines tend to be less competitive in terms of efficiency (and specific fuel consumption) compared with aircraft gas turbines. Against this, high working temperatures can prejudice turbine life and emphasise fuel-quality features to a much greater extent in aircraft-type units. A further performance consideration is the number of rapid starts required of a unit; in a situation where speedy attainment of full load is essential, the aircraft gas turbine can achieve this condition in 1½ minutes, or at least three times more quickly than is usual with the heavy industrial machine. Hence, in an application such as electrical generation for 'lopping' peak demand, the aircraft-type gas turbine has the more favourable start-up characteristics, but frequent starts can be detrimental to turbine life. However, it is forecast that these potential disadvantages will be outweighed by the thermal efficiency benefit, resulting in continued emphasis on modified aircraft gas turbines in many applications.

FUELS

Fuel quality is, and will continue to be, a significant factor in industrial gas turbine operation, though because of the negligible contact between lubricant and combustion products, fuel quality by itself does not have such an important influence on lubricant-quality requirements, as in, for example, the diesel engine. Fuels for industrial gas turbines can range from natural gas to residual fuel and, in the past, low thermal efficiency could be offset, at least in part, by the use of cheap fuel. This can still be so, but the use of the highly efficient aircraft-type gas turbine is highlighting fuel-quality features, and making residual fuel operation less attractive. Natural gas, butane, naphthas, and middle and heavy distillates are all potential fuels, but factors such as stringent limitations on vanadium and sodium contents will place greater emphasis on the use of naphthas or gases in aircraft-type turbines, while manufacturers' ash and viscosity restrictions will limit the fuel used in heavy industrial units. While the choice of fuel will be influenced by circumstances, the use of natural gas can give a significant extension in overhaul turbine life compared with distillate fuel. This is particularly so in the modified aircraft-type turbine, where improvement can be three-fold in a 'fast-start' regime. This figure is not so dramatic for the heavy industrial unit, but up to 25 per cent improvement can be achieved by moving from distillate to natural gas.

LUBRICANT REQUIREMENTS

Lubricating oil in a gas turbine performs the normal function of preventing wear by creating a liquid film between moving parts, while it must also act as a coolant to remove heat from critical bearing areas. Components to be lubricated can include main bearings in the gas generator and power turbine systems, reduction gearing and gearing for auxiliary drives and control systems. Even though oil make-up is usually very small (for example, 6×10^{-7} kg/kWh) the demands on the lubricant are, in many respects, less severe than in the case of a diesel engine. For example, there are no reciprocating parts to create critical lubrication conditions, nor is the oil exposed to contamination by combustion products, which include strong acids, fuel oxidation products, carbon and moisture.

The main requirement for a gas turbine lubricant is for a satisfactory level of oxidation and thermal stability to enable the oil to remain in use for a reasonable period of time without significant increase in viscosity, without the formation of insoluble degradation products that could promote oil system malfunctions, and without deposition of lacquer and carbon deposits in the main bearing area.

The actual level of stability required varies widely with different types of gas turbine, and it is convenient to consider separately the heavy-duty gas turbine and the modified aircraft gas turbine.

Heavy-duty gas turbine

A typical lubricating oil system for this type of unit is illustrated in figure 5.3. The quantity of oil in the system is relatively large. For example a 20 MW power unit might have an oil charge of 7000 litres (compared with perhaps 100–140 litres for the gasifier section of a modified aircraft-type unit with the same power output). In the case of the heavy-duty unit, the common lubricant system supplies oil to the gasifier section bearings and power turbine bearings, and any reduction gearing and control equipment.

With the heavy-duty unit, which has fairly conservative values for pressure ratio and turbine inlet temperatures, and when provision can be made for heat shielding and cooling of critical parts, the temperatures in lubricated areas can be maintained at moderate levels. For example, the oil supply to the bearings would be in the range 50–70 °C, with a rise of 15–30 °C across the bearings, and with a bulk oil temperature in the tank in the range 65–95 °C.

Most heavy-duty types of gas turbine can be satisfactorily lubricated with petroleum-based products of the types used as steam turbine lubricants or hydraulic oils. These are generally formulated with high viscosity index (95–105) solvent-refined base stocks of low viscosity, to provide the required high-temperature oxidation stability and the minimum variation in viscosity with temperature. Oxidation stability is further enhanced by oxidation inhibitors, and

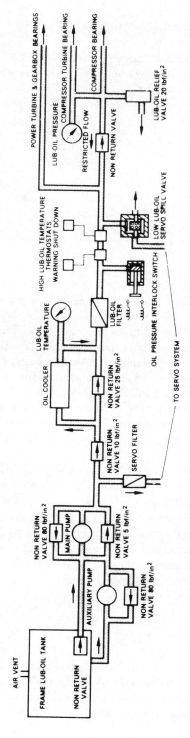

Figure 5.3 Lubricating oil supply systems for Ruston TA gas turbine (courtesy of Ruston Gas Turbines Ltd)

Table 5.1 Typical characteristics of a steam turbine lubricant
and a hydraulic oil for heavy-duty industrial gas turbines

	Steam turbine oil	Hydraulic oil
Density at 15.5 °C	0.861	0.872
Viscosity at 40 °C (cSt)	29.5	30.6
Viscosity at 100 °C (cSt)	5.20	5.35
Viscosity Index (extended)	106	108
Flash point (COC) (°C)	232	210
Pour point (°C)	−12	−30

frequently by compounds designed to reduce the catalytic effect of some metals on oxidation. A rust-inhibited additive is included to protect the lubricant system from rusting.

Some gas turbine designs subject the lubricant to higher temperatures, which make typical steam turbine lubricants unsuitable. These units require special lubricants of increased oxidation stability. In certain cases, for instance where reduction gearing is incorporated, an EP-type additive may be necessary. Pour depressant and anti-foam additives will also commonly be included.

The normal range of viscosity for gas turbine lubricants is 32–68 cSt at 40 °C. Typical inspection characteristics for a steam turbine lubricant and a hydraulic oil used for the lubrication of heavy-duty industrial gas turbines are summarised in Table 5.1.

Although consumption of lubricant in a heavy-duty industrial gas turbine is normally negligible, with very small amounts of fresh oil make-up, the rate of deterioration of oil is normally very slight, so that a typical oil charge would not require draining and replacement for 15–20 years. When monitoring of used oil condition is required, sampling on an annual basis would normally be adequate, unless an unusually severe operation requires more frequent sampling. Used oil analysis would normally include viscosity, neutralisation number, moisture content and insolubles-content determination. Other tests for rust prevention or oxidation stability may also be included.

Modified aircraft gas turbine

The aircraft-type gas turbine units (which function as gas generators for the specially developed power turbine section) were developed originally as high-output compact units with high pressure ratios and turbine inlet temperatures. As such they need much greater levels of oxidation and thermal stability than are obtainable with petroleum-based lubricants, and require the same synthetic lubricants as developed for aviation application.

The power turbine is normally lubricated by a separate system using a petroleum-based lubricant (steam turbine oil or hydraulic oil) as discussed in the previous section on the heavy-duty gas turbine. The same petroleum-based lubricant may also be used for lubricating the driven machine, which in the case of electricity generation would involve the alternator and exciter bearings, auxiliaries and the control system. The quantity of oil in the system is relatively large, typically 4000–7000 litres for a 25 MW machine.

The gasifier section has a separate lubricant system for the synthetic oil that lubricates the compressor and compressor turbine bearings, and a typical arrangement is shown in figure 5.4. In this case the quantity of oil in the system is much less than for the power turbine system, or for a heavy-duty gas turbine of equivalent power. For example, a 25 MW machine may have a total lubricant capacity of only 100–140 litres, which in fact represents a considerably increased lubricant capacity compared with the aviation application, to allow for unattended running.

Operating temperatures in the lubricant system of the gasifier section are considerably higher than for the power turbine or heavy-duty gas turbine systems. Bulk oil temperatures may be in the range 120–150 °C and the oil temperature from the bearings may exceed 200 °C.

As regards synthetic lubricants for the gasifier section, a number of types are available from the aviation field. All of them use ester base stocks along with appropriate additives to achieve the desired performance characteristics. Esters are the reaction product between an organic acid and an alcohol, with the elimination of water, analogous to the formation of an inorganic salt from an inorganic acid and an inorganic base.

Synthetic aviation lubricants of the type developed in the 1950s, now known as Type 1, have nominal viscosities at 100 °C of either 3 cSt (meeting the US specification MIL-L-7808) or 7.5 cSt (meeting the UK specification D. Eng. R.D. 2487). The 3 cSt lubricants were designed specifically for turbojet engines, whereas the 7.5 cSt lubricants were developed mainly for turboprop engines to cater for the highly loaded reduction gearing. Both types are based on diesters of the following arrangement, formed from a difunctional organic acid and a monohydric alcohol:

Alcohol—Dibasic acid—Alcohol

The higher viscosity 7.5 cSt products additionally contain a thickener, which may be of a polymeric nature (such as polyglycol ether) or a complex ester of the type:

Alcohol—Dibasic acid—Dihydric alcohol—Dibasic acid—Alcohol

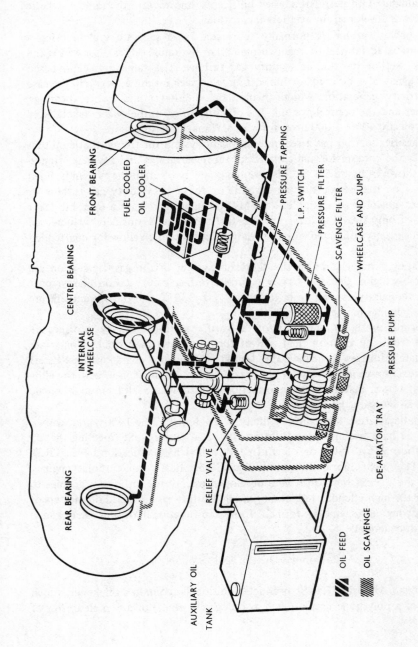

Figure 5.4   Avon gas generator Mk 1533 lubrication system, showing engine-
mounted fuel-cooled oil cooler
(courtesy of GEC Gas Turbines Ltd)

Both types of lubricant contain an anti-oxidant additive, and the excellent high-temperature performance of these ester lubricants is due largely to their marked response to this class of additive. The formulations may also contain anti-foam additive (a few parts per million of a silicone polymer) and they may also contain copper passivators and load-carrying agents.

Type I diester-based lubricants still represent a significant proportion of the market for synthetic aviation lubricants. However, many modern turbojet and turbofan engines require levels of thermal and oxidative stability exceeding the quality provided by Type I products, and these have been met by the development of second-generation lubricants, known as Type II products. These are based on polyfunctional alcohols esterified with monobasic acids, giving the so-called 'hindered' ester structure:

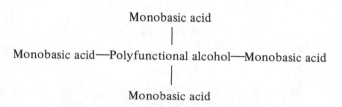

This type of ester gives roughly a 50 °C advantage in thermal stability over the diester type of molecule. The reactions by which a diester decomposes are inhibited by the particular structure of the 'hindered' ester, and a different mechanism is involved, which requires greater energy and which can only occur at higher temperatures.

The Type II lubricants generally have a 100 °C viscosity around 5 cSt, compared with the 3 cSt and 7.5 cSt levels of Type I products, but there are some Type II products of 3 cSt. There are also now on the market improved synthetic lubricants that exceed Type II in performance. They are designed to meet the demands of supersonic aircraft and to give improved performance in lower-rated engines.

Typical characteristics of these various types of synthetic aviation lubricants are summarised in Table 5.2 in comparison with mineral oils.

Table 5.2 illustrates the three main advantages for ester lubricants. Firstly their high natural viscosity index ensures that low-temperature properties, for a given working viscosity, are markedly superior to those obtained from mineral oils. Secondly, volatility characteristics for esters are lower than for mineral oils of the same viscosity level. Thirdly, the load-carrying ability of an ester is significantly better than that of a mineral oil of similar viscosity; this is attributed to the very effective boundary lubrication characteristics of the polar ester structure.

Engine manufacturers' recommendations should be consulted in deciding on the particular type of synthetic lubricant to be used in a specific gas turbine

Table 5.2 Summary of typical characteristics of various synthetic aviation lubricants compared to those of mineral oils

	Petroleum-based lubricants		Synthetic lubricants			
			Type I		Type II	
Nominal Viscosity at 100 °C (cSt)	2.5	20	3	7.5	3	5
K.V. at 100 °C (cSt)	2.5	20.1	3.5	7.57	3.26	5.09
K.V. at 38 °C (cSt)	10.09	266.5	13.9	36.0	13.46	26.3
K.V. at −40 °C (cSt)	3000	Solid	—	11000	—	7760
K.V. at −54 °C (cSt)	25000	Solid	12300	—	12716	—
Viscosity Index (Extended)	75	95	148	199	122	135
Pour point (°C)	<−60	−18	<−60	−51	<−57	<−57
Vapour pressure (Pa)						
204 °C	24120	<7	110	97	193	0
288 °C	—	448	3586	1586	2759	690
Evaporation loss (per cent), 6½ hours at 204 °C	—	—	10	11	17	4.2
Ryder gear test load (lb/in.)	600	Ca 3000	2300	3200	2575	2796

engine. As in the case of the heavy-duty type industrial gas turbine unit, oil consumption characteristics are negligible. Deterioration of the lubricating oil charge in service is normally very slight and a complete change of oil would not normally be carried out between major engine overhauls. Periodic monitoring of used oil condition is normally assessed by means of viscosity and total acid number checks.

Petroleum and synthetic oils are not interchangeable. The use of a synthetic oil in a unit designed for a petroleum oil may be expected to cause difficulties especially with lubricant seal incompatibility. Also synthetic oils can adversely affect paints, insulating materials and many plastics that are perfectly suitable for use with petroleum lubricants. Materials compatible with synthetic oils are, however, available in all areas.

6 Lubrication of Gears and Drives

M. G. Hayler *Ph.D., B.Sc., C.Eng., M.I.Mech.E.*
Esso Petroleum Company Limited
S. J. Crampton *B.Sc., C.Eng., M.I.Mech.E.*

The essential factor in the lubrication of gearing, as with other mechanisms, is the maintenance of a fluid film between the surfaces of components with relative motion and mutual loading. By separating the surfaces in this way, there is a reduction in the degree of contact between the asperities on the mating parts. Inadequate separation leads to welding of the areas of contact, followed by tearing (scuffing) of the metal surfaces and the virtual destruction of the gears.

The degree of contact of the mating surfaces is governed by the loading, by the surface finish, and by the effective viscosity of the lubricant under the local conditions of temperature and pressure.

At one extreme, the load is supported almost entirely by metal-to-metal contact, with the minimum of pressure in the intermittent oil film. Under these conditions of boundary lubrication, wear and scuffing may be so severe that the surfaces of the gear teeth are irreparably damaged.

At the other end of the scale, full hydrodynamic lubrication will support the load entirely on the oil film, and contact of the asperities will be virtually non-existent.

In practice, gearing will most often operate under conditions between these two extremes. That is to say, the force between the teeth will be carried in varying proportions by opposing asperities and by the fluid film. Provided that that film is adequate when the gears are new, then the running-in process will reduce further the metal-to-metal contact and increase the proportion of the load taken by the oil film.

It follows that choosing a lubricant for a given duty becomes a matter of assesing the lubricant properties needed to provide adequate film thickness under those operating conditions. Thus, the required minimum film thickness will be governed by the surface roughness of the conjoined surfaces, that is, the film thickness should exceed the combined heights of the asperities on mating teeth. The actual thickness of the film generated will depend on the relative velocity of those surfaces, their dimensions, and the viscosity of the oil under those conditions.

Therefore, a lubricant is selected with a viscosity that will permit the necessary film thickness to give a predominantly hydrodynamic mode of lubrication.

The higher the asperities (that is, the rougher the surfaces), the greater needs to be the film thickness to separate them. Conversely, the smoother the surface, the less will be the degree of contact, and the film may be correspondingly thinner. Hence, any method of improving the surface finish of mating components will reduce wear rates, for a lubricant of given viscosity, or permit the use of a lubricant of lower viscosity.

Surface finish, or more accurately surface roughness, is dependent, within limits, on the method of manufacture, and the latter is in turn governed by the material of the gear and its application. Thus, lightly loaded spur and helical gears are shaped or hobbed from mild steel blanks, with no further treatment or finish. Gears for heavier duty will be either surface- or through-hardened. If the heat treatment induces unacceptable distortion then the teeth can be shaved or ground.

Simple machine-cut gears will have a surface roughness (CLA) in the range of $1-2.5\,\mu m$. Shaved or ground gears will have a smoother finish, in the $0.13-1\,\mu m$ range. But, in all cases there will be a tendency for the surface to improve in service initially, provided the loading is kept to a reduced level. This phenomenon relies on the controlled reduction of hydrodynamic lubrication, so that the degree of asperity contact is relatively high. The consequent welding and tearing reduces the asperities to a more uniform height, thus enabling a higher proportion of the load to be supported on the oil film.

It follows that the process of running-in is more appropriate to the rougher, softer gears, and they are more susceptible to a period of light-load running. Hardened gears will generate smoother surfaces at a lower rate, so that an initial period of several months' operation may elapse before the running-in can be said to be complete. In some cases the surface condition may change so little in service that the running-in process is imperceptible.

EFFECT OF OPERATING CONDITIONS

That a lubricant relies on viscosity for its ability to lubricate could suggest that the viscosity could never be too high. In practice, of course, factors other than viscosity have some effect on the choice of oil for a given application.

In general, the higher the viscosity the more friction or 'drag' does the lubricant generate. The higher the frictional torque, the greater is the power loss in the gears and the lower is the mechanical efficiency. There is, of course, an economic incentive to reduce the power loss, since both running costs and the cost of initial equipment will be unnecessarily high.

The practical effect of the power loss will be to raise the temperature of the lubricant, thereby reducing the viscosity, until equilibrium conditions occur.

Hence, the higher the viscosity of the oil, the higher the running temperature, with a consequential deleterious effect on the life of the lubricant. This aspect of gear lubrication will be referred to in more detail later.

If high oil temperatures lead to lower viscosity and shorter oil life, there is an incentive to keep the bulk temperature as low as possible. Environmental conditions may or may not be conducive to low temperatures. High ambient temperatures, or the proximity of hot equipment, as in a steam process plant or smelting works, suggest a relatively viscous lubricant. But similar plant may be required to operate in near-arctic conditions, for which a lighter oil would be appropriate, since heat dissipation rates will be higher at the lower temperature.

Not infrequently the piece of equipment will need to start-up at very low ambient temperature and operate subsequently at high temperature. Two possibilities exist: either we select an oil to suit the low-temperature operation, and employ an oil cooler at high temperatures; or the oil is chosen to suit the high-temperature operation and is heated to the appropriate temperature for low ambients.

Which method is adopted depends on the frequency and duration of each mode of operation. If the plant functions predominantly at low temperature, then a light lubricant with a cooler is preferable. If the operation is mainly at high temperature, then a heavier lubricant and a heater could be called for.

Broadly, heating is more easily and economically installed than cooling. So easily, in fact, that care is needed to avoid local overheating by the heating element, be it electrical, steam or hot-water coils. The rate of heat transfer from the element should not be higher than one watt per square centimetre. Higher rates of heat release will tend to degrade the oil by thermal breakdown and oxidation.

The viscosity of a lubricant for a particular application will be defined mainly by the size and shape of the gear teeth and the speed of operation. From these factors the required viscosity can be determined to a first approximation. The value thus indicated is the viscosity required for full hydrodynamic lubrication, and will be unaffected by considerations of temperature. Thus, theory may indicate that a viscosity of 10 cSt is needed in a particular set of gears. If the gears operate at 0 °C then the oil will need a viscosity of 10 cSt at 0 °C, and if the operating temperature is 100 °C then the required viscosity will be 10 cSt at 100 °C.

In connection with low-temperature operation, it should be noted that the pour point test (IP 15) may give an optimistic indication of the temperature at which the lubricant ceases to flow. That is to say, IP 15 may give a pour point that is considerably lower than the operational limit in practice, the actual difference depending on the lubrication system employed.

With splash lubrication, where one or more gears dip into the oil, a degree of 'oiliness' will exist during the critical start-up period while the oil is cold, although the oil may tend to 'channel', with subsequent starvation.

When a pumped spray system is used, the limit of low-temperature operation, with a particular lubricant, will depend on the position of the pump. If the pump is submerged below the oil level, then oil will flow to the sprays more readily than if the pump has a 'lift' on the suction side.

But in all cases, oil circulation, by whatever means, can be restricted on cold starting at temperatures as high as 20 °C above the indicated pour point. Where possible, loading should be reduced until the oil attains a temperature that permits full circulation. If no reduction of load is possible, oil heating may be necessary.

Discussion so far has centred on those properties needed to ensure the generation and maintenance of an adequate lubricant film thickness under the conditions that the gears are likely to encounter in service. Other properties, depending on the application of a particular oil, are also necessary.

In service, an oil will tend to oxidise and acidify, thereby increasing both its viscosity and its corrosive tendencies. Any increase in temperature will aggravate the process of oxidation, so that, unless oil changes are frequent, or the system is total loss, the temperature of the lubricant should be kept as low as is practicable.

For a longer interval between oil changes, the rate of oxidation can be reduced by the use of additives. Suitably inhibited, the life of the oil can be extended considerably, and with appropriately mild conditions and a suitably high rate of top-up, a lubricant can outlast the equipment in which it serves. Conversely, an oil that lubricates on a 'once-through', or total loss, system need not incur the cost of oxidation inhibition, and may be of lower inherent quality.

Corrosion may be a potential problem in some applications. In particular, the oil may attack yellow metals, such as copper alloys, as used for bearings, pipes and fittings, worm-wheels etc. Increase of acidity with ageing of the oil can increase the degree of staining and corrosion.

Similarly, rust protection of the gears can be increased by the addition of compounds to the oil. Such properties are of benefit if the gears run only intermittently, particularly in humid atmospheric conditions or when the oil itself becomes 'wet' from contact with water or steam.

If contamination by water is liable to occur in a particular system, then the ability of the oil to shed water must be evaluated and, if necessary, augmented. Water will tend to drop out of the oil if the circulation rate is low enough. If, on the other hand, churning of the oil is continuous, as in a splash-lubricated unit, the oil and water may form a fairly permanent emulsion, and water separation may not occur naturally. The addition of a demulsifying agent will improve the water-shedding properties, but if large volumes of oil are involved, then centrifuging becomes a feasible proposition.

TYPES OF GEAR LUBRICANT

Lubricant viscosity governs the film thickness under elasto-hydrodynamic conditions. However, there is a practical limit to the viscosity from other considerations. Hence, there will be gear units where the loading is such that the full film thickness cannot be maintained, so that boundary lubrication conditions predominate. High wear, scuffing and scoring would normally be the consequences of continuous boundary lubrication, but it is possible to mitigate these effects in varying degrees, by the use of additives in the oil.

These additives operate in different ways, but can be regarded, in effect, as artificial extensions of the viscosity scale; they provide the benefits of higher viscosity without the disadvantages that unduly high viscosity brings. They can, however, bring their own problems of other sorts.

The most simple type of additive to enhance load carrying is the 'oiliness' additive, or 'film-strength improver'. The method by which they operate is not well understood, but it is accepted that there is no chemical action involved. By attachment to the surfaces they resist the squeezing action that tends to exude the oil film, and thereby reduce the effects of contact of the asperities.

The so-called 'compounded oils' contain, usually, 3–8 per cent of fatty oil, frequently acidless tallow. The consequent reduction in friction is frequently exploited in worm gears, where the high sliding velocity generates considerable heat.

Additives of these types must not be confused with the so-called 'extreme pressure', or EP, additives, which operate by chemical combination with the surface of the mating components. These additives usually contain sulphur, phosphorus or chlorine, or combinations of those elements, which react under conditions of high temperature and high pressure. The compounds thus formed behave as a lubricant film, separating the asperities and preventing their welding and tearing. By this means additional load-carrying properties can be built into the oil without the attendant disadvantages of high viscosity.

Consideration of the mode of action of EP additives suggests that EP activity will tend to remove the outstanding metal and leave a smoother, more polished surface. Observations of oils of this type in service confirm the tendency for the surface finish to improve during the initial phase of the operation. It follows that the normal, physical, mode of lubrication will become more appropriate as this period of running-in proceeds, so that EP activity is needed less, and may, in some cases, be dispensed with, the EP oil being replaced by a straight mineral oil.

The effectiveness, or activity, of EP additives varies with the type of additive. The less powerful types are now generally known by the somewhat contradictory description of 'mild EP' additives. Of these mild EP types, the most gentle have been referred to as 'anti-wear' agents, of which the zinc dithiophosphates and tricresyl phosphate are probably the most widely used.

Anti-wear agents and mild EP additives can be thought of as functioning to increase the load-carrying area of the gear teeth by smoothing or swaging the asperities, whereas the 'full' EP additives achieve a similar effect (and carry more load while doing it) by chemical removal of the asperities. It will thus be apparent that EP additives are, strictly, 'pro-wear' additives, that is, they function by removing metal. However, as the surface finish improves the high local temperatures that trigger the action of the EP additive become less common. The activity therefore decreases, as does the need for it.

EP additives will be essential not only when the loading on the teeth is continuously high, but also when the loading is suddenly imposed. Roll drives in steel mills are an example. The load on the gears during rolling can be assessed, and the teeth so designed and stressed that an EP oil would not normally be needed. But it is not feasible to apply the load gradually. Nor is it economic to design the gears so that the shock load would be the normal working load of the gears. It is, therefore, usual to use an EP oil with an inbuilt protection for shock loads. The traditional additive for this type of service was lead naphthenate, augmented by a sulphurised fatty oil, but this has been replaced in many plants by sulphur/phosphorus additives.

For open gears, particularly those operating in external conditions, oils compounded with bitumen are necessary to adhere to the gear teeth. These lubricants are applied either hot or cut-back with white spirit to a suitable viscosity. The solvent evaporates, leaving the oil adhering to the gears.

GEAR LUBRICANT SELECTION

In general, gearing will be dimensioned according to the loads and speeds under which it has to operate. Considerations of elasto-hydrodynamic lubrication theory indicate that those two factors work in opposition in the determination of the appropriate viscosity for the necessary thickness of the lubricant film. Thus, the higher the loading between the teeth, the higher the viscosity of the oil that is needed, and the higher the speed of operating, the lower the viscosity.

In practical industrial applications, the majority of gearing is of the straight spur (figure 6.1) or the helical types. The latter is more expensive to produce but is quieter in service and provides a larger tooth area for a given face width. Thus, for a gear that needs to transmit a given torque, the tooth stresses will be less with helical gears than with spur gears, and a lighter, less viscous, grade of lubricant will suffice. But, in general, spur and helical gears can be adequately lubricated with straight oils. Load-carrying additives will only be necessary if the gears are loaded beyond their design capacity, or if circumstances dictate the use of a lighter grade of oil than operating conditions would otherwise suggest.

Bevel gears, be they straight or spiral, will have requirements similar to those of spur gears for lubricants. Two factors may, however, influence the selection. Firstly, the angle between the axes of the gears needs to be accurately maintained, so that rolling bearings are normally used with this type of gearing. Secondly, the axial thrust, particularly with the spiral bevel type, may adversely affect the operation of the bearings, so that bearing lubrication may be the governing factor in the selection of a lubricant.

Generically, the hypoid gear falls between the spiral bevel and the worm gear. The off-set of the pinion axis produces a mode of operation with a high slide-to-

Figure 6.1   Straight spur type of industrial gearing

roll ratio, the degree of sliding increasing with the amount of off-set. Further, tooth stresses are high, so that both gears are made in steel. This combination of high loading and high rubbing velocity calls for a 'full' EP oil, particularly during the breaking-in period. 'Mild' EP oils may be adequate thereafter, but EP lubricants will be needed throughout the working life of the gear, if scuffing is to be avoided.

Worm gears have the distinction that the relative motion between the worm and wheel is virtually all sliding, and thereby generates considerable amounts of heat. The lubricant will assist in the transmission of this heat from the gears to the casing, from which it will be dissipated into the atmosphere. In most cases, the power transmitted by a worm gear is limited by this temperature rise, rather than by the physical loading of the gears. EP additives are superfluous, and may, if too active, cause staining or corrosion of the worm wheel, which is usually of bronze. More important are resistance to oxidation and to thermal breakdown, and worm gear lubricants are usually straight mineral oils, with the addition of compounds to resist deterioration and to inhibit rusting and corrosion.

The so-called 'compounded oils' can be beneficial for worm gears, as the addition of fatty oils to a mineral base will reduce the coefficient of friction, and thereby limit the temperature rise. The gear unit can therefore transmit higher power, but the fatty oils tend to be inferior in thermal stability, and require more frequent oil changes.

Also effective in reducing friction are the polyglycol ethers, the 'synthetic' lubricants. These compounds have high thermal and oxidation stability, high viscosity index and good low-temperature fluidity. Their stability is such that the initial charge of lubricant will generally outlast the life of the gears. The main disadvantages of these compounds are their price (about three times that of mineral oil), and their searching properties, which make sealing a problem. The most minute gap or clearance will permit seepage. The effects of this searching can be reduced by thickening the polyglycol ether, usually with lithium hydroxystearate, to form a grease. Even then, 'bleeding' of the polyglycol ether can take place, giving a free liquid which is as prone to seepage as the 'straight' polyglycol ether.

These synthetic greases are most commonly used in smaller worm gear units — up to 150-mm centres — where the temperature rise is less, owing to the more favourable surface-to-volume ratio (the static grease relies solely on conductance for heat transfer to the gear casing, convection being absent). In the larger gear units, an oil-circulating system with sprays will ensure the maximum heat transfer and cool running.

The American Gear Manufacturers' Association (AGMA) has defined a range of industrial gear lubricants. Three types of oil are envisaged, R and O (rust and oxidation inhibited), compounded, and EP, the compounded oils being an extention of the R and O category. Oils of each type are numbered according to viscosity, as shown in Tables 6.1 and 6.2.

Having defined the grades of lubricant, AGMA also makes recommendations as to the selection of oils for particular applications. These recommendations are shown in Tables 6.3, 6.4 and 6.5.

It must be emphasised that these recommendations are general, whereas almost every application has its particular conditions. Nevertheless, the AGMA recommendations indicate an acceptable starting point in the process of lubricant selection, and a grade ascertained from the tables can be varied to accommodate these conditions. Thus, for heavily loaded gears, or high operating temperatures, a grade heavier than the tables indicate would be advantageous. Low-temperature operation and high speeds call for a less viscous oil than the standard recommendation.

Table 6.1   Viscosity range for AGMA R and O and compounded lubricants

AGMA number	Viscosity range (cSt) at 40 °C
1	41.4-50.6
2	61.2-74.8
3	90-110
4	135-165
5	198-242
6	288-352
7 comp[a]	414-506
8 comp[a]	612-748
8A comp[a]	900-1100

[a] To be compounded with 3 - 10 per cent animal fat such as acidless tallow.

Table 6.2   Viscosity range of AGMA EP lubricants (FZG Load Stage 11 or Timken OK load of 60 lb minimum)

AGMA EP lubricant number	Viscosity range (cSt) at 40 °C
2 EP	61.2-74.8
3 EP	90-110
4 EP	135-165
5 EP	198-242
6 EP	288-352
7 EP	414-506
8 EP	612-748
8A EP	900-1100

Table 6.3 AGMA recommendations for lubrication of enclosed gearing

| Main-gear low-speed centres | | Ambient temperature (°C) | |
Type of unit	Size of unit (mm)	−10 to 10 Use AGMA number	10 to 50 Use AGMA number
Parallel shaft (single reduction)	Up to 200	2–3	3–4
	Over 200 and up to 500	2–3	4–5
	Over 500	3–4	4–5
Parallel shaft (double reduction)	Up to 200	2–3	3–4
	Over 200	3–4	4–5
Parallel shaft (triple reduction)	Up to 200	2–3	3–4
	Over 200 and up to 500	3–4	4–5
	Over 500	4–5	5–6
Planetary gear units	OD housing up to 400	2–3	3–4
	OD housing over 400	3–4	4–5
Spiral or straight bevel gear units	Cone distance up to 300	2–3	4–5
	Cone distance over 300	3–4	5–6
Gear motors	—	2–3	4–5
High-speed units[a]	—	1	2

[a]For speeds over 3600 rpm or pitch-line velocities over 25 m/s.

Table 6.4 Lubricants for cylindrical and double-enveloping worm-gear units (AGMA 250.04 – September 1981)

Worm centres	Worm speed up to (rpm)	Ambient temperature (°C)		Worm speed above (rpm)	Ambient temperature (°C)	
		– 10 to 10	10 to 50		– 10 to 10	10 to 50
Up to 150 mm inclusive:						
Cylindrical worms	700	7 comp, 7EP	8 comp, 8EP	700	7 comp, 7EP	8 comp, 8EP
Double-enveloping worms	700	8 comp	8A comp	700	8 comp	8 comp
Over 150 mm centres up to 300 mm centres:						
Cylindrical worms	450	7 comp, 7EP	8 comp, 8EP	450	7 comp, 7EP	7 comp, 7EP
Double-enveloping worms	450	8 comp	8A comp	450	8 comp	8 comp
Over 300 mm centres up to 450 mm centres:						
Cylindrical worms	300	7 comp, 7EP	8 comp, 8EP	300	7 comp, 7EP	7 comp, 7EP
Double-enveloping worms	300	8 comp	8A comp	300	8 comp	8 comp
Over 450 mm centres up to 600 mm centres:						
Cylindrical worms	250	7 comp, 7EP	8 comp, 8EP	250	7 comp, 7EP	7 comp, 7EP
Double-enveloping worms	250	8 comp	8A comp	250	8 comp	8 comp
Over 600 centres:						
Cylindrical worms	200	7 comp, 7EP	8 comp, 8EP	200	7 comp, 7EP	7 comp, 7EP
Double-enveloping worms	200	8 comp	8A comp	200	8 comp	8 comp

Table 6.5 Recommended AGMA EP lubricants for industrial enclosed gearing

Type of unit	Size of unit main-gear low-speed centres (mm)	AGMA EP lubricant number ambient temperature (°C)	
		−10 to 10	10 to 50
Parallel shaft, single reduction	Up to 200	2 EP – 3 EP	3 EP – 4 EP
	Over 200 up to 500	2 EP – 3 EP	4 EP – 5 EP
	Over 500	3 EP – 4 EP	4 EP – 5 EP
Parallel shaft, double reduction	Up to 200	2 EP – 3 EP	3 EP – 4 EP
	Over 200	3 EP – 4 EP	4 EP – 5 EP
Parallel shaft, triple reduction	Up to 200	2 EP – 3 EP	3 EP – 4 EP
	Over 200 up to 500	4 EP – 5 EP	4 EP – 5 EP
	Over 500	4 EP – 5 EP	5 EP – 6 EP
Planetary gear	OD of housing up to 400	2 EP – 3 EP	3 EP – 4 EP
	OD of housing over 400	3 EP – 4 EP	4 EP – 5 EP
Gear motors	All sizes	2 EP – 3 EP	4 EP – 5 EP
Spiral or straight bevel gear	Cone distance up to 300	2 EP – 3 EP	4 EP – 5 EP
	Cone distance over 300	3 EP – 4 EP	5 EP – 6 EP

Note also that some of the EP oils, notably those formulated with lead naphthenate, and the compounded oils tend to be unstable at high temperatures, and are therefore limited to applications where the bulk oil temperature is less than about 70 °C. If an EP oil is needed for temperatures above that value, then an oil of the 'full' EP type, such as sulphur/phosphorus, may be used.

MODES OF FAILURE

It has already been shown that the essence of gear lubrication is to keep apart, so far as is practicable, the conforming surfaces of mating gear teeth. Since this objective will never be completely achieved, the extent to which lubrication falls short of the ideal will be reflected in the type and degree of damage to the teeth.

Gear teeth will always wear in service, and the best that can be hoped for is the so-called 'normal' wear. That term is imprecise, and normally will vary widely with different types of gearing and different applications. But it can be taken as meaning a rate of wear low enough to give an acceptable life, with no likelihood of catastrophic wear.

Abrasive wear is a term applied to the continuous removal of the surface of the teeth at a rate significantly higher than that of normal wear. Hard, solid contaminants in the oil produce this type of wear by scratching and scoring the surface. Abrasion by fine particles can give an apparently smooth surface finish.

The degree of surface contact for abrasive wear is generally insufficient to produce scuffing, that is, the welding and subsequent tearing out of the asperities. Scuffing leaves a very rough surface, and involves a high rate of metal removal from the teeth. It is an indication that lubrication has failed, and the only remedy that can be applied is to replace the damaged gears. As discussed earlier, the use of an EP additive in the lubricant can prevent scuffing; and hardening of the gears, either through-hardening or surface-hardening, can assist in withstanding high inter-tooth loading.

Scuffing invariably begins at the extremes of the tooth (that is, at the tips and roots) and progresses towards the pitch line. Sliding velocities are highest at those points, so that contact temperatures are highest. At the pitch line, the relative motion of the teeth in contact is pure rolling, so that scuffing is rare, even with gross overloading.

When the tooth loading becomes excessive, a frequent manifestation is the occurrence of pitting. Because the contact at the pitch line is pure rolling, pitting will tend to break out at that part of the tooth, and frequently before the overloading causes scuffing on the addendum and dedendum. If the teeth are hardened, and an EP oil is used to reduce the tendency to scuffing, pitting may occur on the addendum and dedendum at a load lower than that at which scuffing occurs.

Pitting is a fatigue phenomenon, occurring on the surface of material subjected to repeated stress cycles. The sequence of events begins with the initiation of surface cracks. The lubricant aggravates the defect by generating pressure in the crack as the surfaces roll over one another under load. The initial cracks occur by deformation of the surface under the rolling action. Hence, local stresses at the surface are likely to be higher during the running-in period, and if that process is prolonged it may be overtaken by pitting of the pitch line. Having commenced in this fashion, the pitting will either 'heal' by means of a swaging or abrading process, or, if the surface stress is high enough, it will spread to the dedendum and addendum. In the latter case, the pitting is known as 'progressive', but it is usually not possible to predict from the initial pitting whether it will be progressive or not. Reducing the loading, where possible, will delay the onset of pitting, as will, to some extent, increasing the viscosity of the oil, but the effect of the former will be by far the greater.

Pitting of worm wheels is a fairly common phenomenon, with the usual combination of steel worm and bronze wheel. It does not usually affect the running.

TEST METHODS AND SPECIFICATIONS FOR GEAR OILS

In so far as gear lubricants are most frequently based on petroleum oils, the normal methods of testing petroleum products will apply. Viscosity, corrosivity, rust prevention, oxidation stability, etc. are all as relevant to gear lubricants as to other oils.

Tests that are specific to gear oils usually evaluate the properties of load-carrying and EP activity. They come in two types, in one of which fully formed gears operate in mesh under load. For the other type, sliding, loaded contact between suitably shaped pieces of steel simulates the action of gear teeth to a greater or lesser extent, and compares oils in terms of the load carried up to the onset of scuffing.

Of the gear-rigs the three most commonly used are the IAE (UK), Ryder (USA) and FZG (Germany). The FZG is now the most widely used in Europe (figure 6.2). All work on the 'four-square' principle, with a variable, measured torque 'built-in'. They rate oils in terms of the torque transmitted before scuffing takes place. The FZG gears have a long and short addendum, to increase the sliding velocity. These three rigs do not correlate particularly well.

Similarly, the bench tests give varying indications of EP activity. In the UK probably the most frequently used are the 4-Ball EP tester, the Timken EP test and to a limited extent, the Falex EP test.

In application, gear manufacturers decide which of these tests is most appropriate to service in the gear units they manufacture, and call for that method, with appropriate limits, in a specification. But it must be emphasised that all of these tests are empirical, and none is capable of defining absolutely the quality of an oil. Experience of the test will indicate the level of result that will give adequate service in practical applications.

Specifications for gear oils are mostly confined to automotive applications, although some gear manufacturers, and the larger user organisations, such as the

Figure 6.2   FZG gear lubricant tester used to evaluate
the load-carrying capacity of gear oils under
controlled conditions

nationalised industries, produce specifications for industrial gear lubricants. These
latter generally specify the oil by type (that is, straight mineral, mild EP, full EP
etc.), by viscosity, and by load carrying, using one of the tests mentioned
previously. It is usual for a manufacturer to recommend for his products oils that
are available from one or more oil companies. These recommended lubricants
may not all meet any one specification, but they will all have been passed as
satisfactory by the equipment manufacturers, on the basis either of testing or of
practical experience.

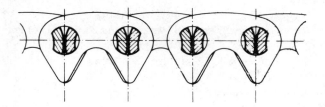

Figure 6.3   Morse 'Hy-Vo' involute chain

OTHER TYPES OF DRIVE

Chains

The chain is probably the most widely used alternative to gears as a means of transmitting power and varying the speed and torque. All types of chain are comprised of the link, pin and bush as the basic elements with, in the case of the inverted-tooth type, an involute tooth operating on the conforming sprocket tooth, instead of the normal roller (figure 6.3). Chain lubrication, therefore, becomes a matter of lubricating a plain bearing and a rolling/sliding element, the materials being steel on steel in each case, And although the speed of the chain as a whole may be high (over 15 m/s), the speeds of the various elements, relative to each other, are low and the type of lubricant is not critical.

The choice of lubricant for chain drives depends on the method of application of the oil, and on the conditions under which the chain operates. At normal temperatures and under the maximum loading recommended by the manufacturer, a straight mineral oil gives adequate lubrication, the viscosity ranging from about 46 cSt to 320 cSt at 40 °C. At low chain speeds (say 2.5 m/s) and on open chains, where hand application is adequate, a relatively heavy oil may be used. At higher speeds, up to 7.5 m/s, drip-feed, bath or splash systems should be employed as a rather more positive means of applying the oil. Up to 15 m/s a slinger can be used to throw oil from the bottom of the oil bath on to the chain, while above that speed oil mist or spray is necessary.

If the chain loading is high an EP gear oil may be necessary, while high temperatures may call for an oil with rust and oxidation inhibition built in.

Special drives

For certain special-purpose drives particular types of lubricant may be needed. Thus, some of the devices giving infinitely variable speed ratios, such as the Kopp Variator, and chain/variable pulley drives, may need oils with special friction characteristics. In these cases only specific lubricants will be recommended by the manufacturer, after considerable testing, and those recommendations should be adhered to.

BIBLIOGRAPHY

I. P. Methods for Analysis and Testing. Volumes 1 and 2, Institute of Petroleum, London, 1982.
Gearing in 1970. Conference, Cambridge, September 1970, *Inst. Mech. Eng. Proc.*, 184 (1969–70) Part 30.
Gear Lubrication Symposium, Brighton, October 1964, Institute of Petroleum, London, 1965.
Performance and Testing of Gear Oil and Transmission Fluids Symposium, Institute of Petroleum, London, October 1980.

7 Hydraulic Transmissions

A. B. Barr *B.Sc., C.Eng., M.I.Mech.E.*
D. F. G. Hampson *C.Eng., M.I.Mech.E.*
Esso Petroleum Company Limited
W. Y. Harper *B.Sc., M.I.Mech.E., A.M.I.R.T.E.*

There are two basic forms of hydraulic power transmission — hydrostatic and hydrodynamic (or hydrokinetic). The hydrostatic drive employs a positive displacement pump and delivers oil to a hydraulic motor, power being transmitted by fluid pressure without great changes in fluid velocity. In the hydrodynamic system a centrifugal-type pump circulates large amounts of oil driving a turbine-type hydraulic motor. Power is transmitted by changes in the fluid velocity.

This chapter will first consider the essential requirements of hydraulic transmission fluids followed by the lubrication requirements of the two power systems.

ESSENTIAL REQUIREMENTS OF HYDRAULIC TRANSMISSION OILS

The characteristics required by an oil suitable for use in a modern hydraulic transmission system can be summarised as shown below:

(1) correct viscosity
(2) resistance to oxidation
(3) non-foaming
(4) good entrained air release
(5) protection against corrosion
(6) adequate low-temperature properties
(7) compatibility with sealing materials
(8) good water separation
(9) anti-wear characteristics
(10) good filterability

To achieve all these requirements in a mineral oil hydraulic fluid even with the use of various additives is not easy and compromises usually have to be made. Even so, today's modern mineral-oil based hydraulic fluids enable hydraulic power systems to operate efficiently for long periods under arduous conditions, with the minimum of attention.

Viscosity

The ideal hydraulic oil would have a constant viscosity at all temperatures, but, unfortunately, it does not exist. The viscosity of mineral oils decreases with temperature rise so that base oils are selected that give the least practical change in viscosity with temperature. In practice good-quality hydraulic oils have a Viscosity Index (VI) of 90–100, which gives acceptable viscosity/temperature characteristics for the majority of indoor and outdoor applications in the UK. The main type of mineral base oil now used in hydraulic oil formulations is termed paraffinic. This relates to its predominant composition characteristics. Paraffinic base oils have a high natural VI (90–110), good rubber swell properties and high oxidation stability, but high pour points (-10 °C). By the use of modern pour depressants the pour point of such oils can be lowered to around -40 °C which is sufficient for most applications.

In any hydraulic system the pump is generally the critical component as regards the viscosity of the hydraulic oil used. All pumps operate most efficiently with a certain viscosity of oil (optimum viscosity). There is also a minimum viscosity and a maximum viscosity of hydraulic oil for each pump. If the oil viscosity is too low, internal leakage within the pump can lead to overheating and mechanical damage. Too high a viscosity can lead to inadequate start-up capabilities in cold weather, and pump cavitation. For these reasons, hydraulic oils are marketed in many viscosity grades and it is important to select an oil of the viscosity range recommended by the pump manufacturer.

Hydraulic oils used in mobile equipment in northern America and Scandinavia have to operate over wide ambient temperature ranges. Special hydraulic oils with a high Viscosity Index (150–200) are used, which give the necessary viscosity at high temperatures but adequate fluidity at low temperatures. In order to obtain such oils, long-chain polymers (VI improvers) are added generally to light paraffinic or naphthenic type base oils. These VI improvers tend to thicken the oil more at high temperatures than low temperatures, thus reducing viscosity/temperature changes, as shown in figure 7.1. High VI hydraulic oils which are thickened by the use of polymers can, however, suffer permanent viscosity loss owing to the shearing of these polymers by hydraulic system components such as relief valves. Some of the polymers used in high VI oils have more resistance to shear than others. To select the most shear-stable polymer it is necessary either to carry out full-scale field trials with the actual hydraulic equipment, measuring the viscosity of the oil being tested before and after the

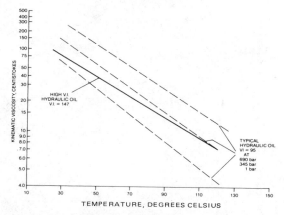

KINEMATIC VISCOSITY, CENTISTOKES

HIGH V.I.
HYDRAULIC OIL
V.I. = 147

TYPICAL
HYDRAULIC OIL
VI = 95
AT
690 bar
345 bar
1 bar

TEMPERATURE, DEGREES CELSIUS

Figure 7.1   Effect of VI improver and pressure on the viscosity/temperature
characteristics of hydraulic oils

test, or use a laboratory rig to simulate the shearing action by the hydraulic
system on the oil. Various rigs have been developed to assess the shear breakdown of high VI hydraulic oils but the one that has gained most acceptance in
Europe is referred to as the Wood's type rig[1]. Shearing of the polymer con-
tained in the high VI oil is achieved by passing the oil through a diesel injector
nozzle by means of a diesel pump for a fixed number of cycles.

There have been many attempts to develop a standard test procedure for this
type of equipment and in the UK the AHEM (Association of Hydraulic Equipment Manufacturers) has played an important role in this respect. Currently
the rig that has gained most acceptance is manufactured in Germany (figure 7.2).
A German standard (DIN 51382) covers the use of this equipment. It was
originally intended for motor oils but now covers high VI hydraulic oils. The
viscosity loss of motor oil as a result of shearing is determined after 30 cycles,
whereas that of high VI hydraulic oils is determined after 250 cycles. The
equivalent Institute of Petroleum test method is IP 294. It is important to
realise that the rig tests themselves can only assess the shear breakdown of one
type of polymer against another. However, after selecting a high VI hydraulic
oil containing a shear-resistant polymer by such rig tests, its actual shear breakdown should be checked in the hydraulic system itself.

The low-temperature viscosity of high VI hydraulic oils should not be extra-
polated from measured viscosities at 40 °C and 100 °C as they can be significantly higher. In practise low-temperature viscosities of high VI oils are usually
determined by a Brookfield viscometer.

Operating pressures used in some hydraulic systems and hydrostatic transmissions are high enough to significantly increase the fluids viscosity, as shown
in figure 7.1. These differences in pressure/viscosity for normal mineral oil can

Figure 7.2   General view of the shear stability rig

be predicted by applying viscosity data at atmospheric pressure to equations
given in reference [2].

Oxidation characteristics

Modern refining methods produce oils with good natural resistance to oxidation.
Generally, the less viscous oils also tend to have greater natural resistance to
oxidation. The oil's natural resistance to oxidation can be extended by the use

of oxidation inhibitors. In general the ideal temperature range for most hydraulic systems is 50–65 °C. However, in some cases hydraulic system temperatures may reach 95 °C and the oil will require changing more frequently than at low operating temperatures.

Oil in a hydraulic system is being constantly circulated and is often subject to agitation in the presence of air. In time the oil thickens as a result of the presence of both soluble and insoluble oxidation products. Other signs of oxidation of a hydraulic oil are an acrid smell, darkening of colour, sludge and lacquer which results in increase of acidity.

The presence of moisture coupled with organic acids may also lead to corrosion of the system components which further adds to sludge formation. Certain metals, especially alloys of copper, can act as catalysts with oils at high temperatures in the presence of oxygen, which can accelerate oxidation of mineral oils. Passivating additives may be incorporated in hydraulic oils to form protective films on the metal surfaces to inhibit this effect.

Foaming

Foam in a hydraulic system can cause a lowering of pump efficiency and loss of control.

Generally, the thinner a hydraulic oil is, the better is its natural resistance to foaming. Foaming of a hydraulic oil is defined as the oil bubbles that collect above the surface of a fluid. The ASTM foam test D 892 measures the tendency for an oil to produce foam and its stability. Usually, it is only necessary to inhibit the more viscous hydraulic oils against foaming by the use of anti-foam additives. One very effective anti-foam additive is silicone which is generally effective when present in only a few parts per million. However, silicone-type anti-foam additives can inhibit the release of entrained air, especially the very small sized bubbles which are sometimes called aero-emulsions. Therefore other types of defoaming additives that do not affect the release of entrained air are generally used in modern hydraulic oils, as shown in figure 7.3.

Air entrainment

All hydraulic oils contain air in solution which has little effect on the operation of the system, although when a hydraulic oil contains air in the form of bubbles in the body of the fluid it can cause problems. Air can be entrained in a hydraulic oil in many ways, usually as the result of a system fault or poor design. Common causes of entrained air are leaks in the pump suction or when the hydraulic return line discharges oil above the surface level of the oil in the reservoir. Increasing attention is being paid to evaluating the ability of an oil to release entrained air and various test methods have been devised to check this characteristic. The test method that has gained most acceptance in Europe is

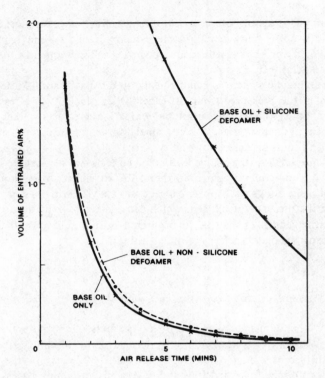

Figure 7.3   Effect of defoamers on entrained air release at 25 °C

DIN 51381 known as the Impinger Method (IP 313).

The effect of entrained air in a hydraulic oil is to increase compressibility. It can also cause cavitation and the collapse of air bubbles on the discharge side of the pump which manifests itself by pump vibrations, noise and sometimes by a destructive effect on metal surfaces (pitting). The natural ability of a mineral oil to release entrained air can be inhibited by silicone defoamers as mentioned previously, and by contamination from surface-active additives such as detergent-dispersants used in engine lubricants.

A recent investigation on aeration in lubricating oils deals more fully with the subject[3]. In this work, a distinction is drawn between lubrication systems in which oil has to separate under tranquil conditions (which benefit from oils without silicone additives), and those in which it has to separate under turbulent conditions (which can benefit from silicone additives). It is suggested, for example, that when oils containing silicone additives are used in hydrostatic drives and by hydrodynamic transmissions using gear pumps, they do not encounter problems of excessive aeration.

Protection against corrosion

Most hydraulic systems contain moisture, especially those used outdoors, owing to the temperature of the air in the oil reservoir falling below its dew point. As most hydraulic systems contain ferrous components it is necessary to prevent rust occurring by the use of additives such as sulphonates and alkenylsuccinic acids, etc. Many laboratory tests exist to evaluate the anti-rust characteristics of hydraulic oils, such as the humidity cabinet rust test and the ASTM D665 method. Other tests are also used to assess any corrosive action on non-ferrous materials used in hydraulic system components.

Low-temperature properties

Hydraulic oils must be able to flow in adequate amounts to fill the displacement volume of the pump at the lowest operating temperature. The temperature at which the oil changes from a fluid to a solid state is designated as the pour point of the oil. Hydraulic oils having pour points of about $-10\ °C$ are satisfactory for use in indoor systems but for outdoor systems the pour point must be at least $10\ °C$ below the lowest expected operating temperature.

Compatibility with sealing materials

Hydraulic fluids must have good compatibility with the seal materials used in the hydraulic system. Small amounts of swelling and shrinkage of fluid seals can be tolerated but excessive changes can affect the seal life and the efficiency of the hydraulic system. The most widely used seals in hydraulic systems today are nitrile rubbers, which have excellent resistance to normal mineral oils. Table 7.1 gives data on these types of seals from a seal manufacturer.

Demulsibility

Water in a hydraulic system must separate rapidly from the oil if the efficiency of the system is to be maintained. The additives in the hydraulic oil should not be types that encourage emulsification, as damage to the pump rolling bearings may result. However, engine oils are often used in hydraulic systems attached to vehicles. These oils contain detergent-dispersant additives, which generally do not have good demulsibility characteristics. Most mobile hydraulic equipment operating out-of-doors runs at bulk oil temperatures of over $65\ °C$, at which temperature the water will evaporate harmlessly.

In addition to detergent-dispersant engine oils there are detergent-type hydraulic oils. These have found acceptance for some machine tool applications, especially in Germany. Their main use is in applications where contamination with aqueous metal-working fluid occurs. In this case it is claimed that it is better to have an emulsion circulating in the hydraulic system than large slugs of water.

Table 7.1 Characteristics of seal material for use in hydraulic systems

Material	Hardness range ($°BS$)^a	Characteristics
Low/medium nitrile	40 – 90	Good resistance to mineral oils. Good to fair high- and low-temperature properties (-50 °C to +100 °C)
Medium/high nitrile	50 – 95	Excellent resistance to mineral oils. Fair to poor at low temperatures. Good high-temperature properties (maximum 130 °C)

^{a} BS Degrees designation for rubber hardness, identical with IRHD (International Rubber Hardness Degrees). Hardness most commonly used for seals is in the range 65–80 °BS.

Anti-wear characteristics

Apart from wear arising from abrasive contaminants entering the hydraulic system, wear also stems from the system itself. Some wear is inescapable, leading to finely divided metallic debris in the hydraulic oil. To minimise such wear an EP-type additive is usually included in a premium hydraulic oil formulation, although many pumps could operate satisfactorily on plain mineral oils. However, there are other pumps, such as the Sperry Vickers Rand Vane Pump (steel on steel), in which wear can be reduced and the pump life extended (figure 7.4), especially when operated under overload conditions by the use of an EP additive.

For many years tri-cresyl phosphate (a mild EP additive) was included in some hydraulic oil formulations, but demand for oils with increased anti-wear characteristics in vane pumps has resulted in other EP additives being used, such as zinc dialkyldithiophosphate (ZDDP). Unfortunately, the metallurgy of some pumps' rubbing components, mainly silver, can be attacked by the ZDDP-type additives in hydraulic oils (figure 7.5). Currently, most premium hydraulic oils contain ZDDP as the anti-wear additive, but other hydraulic oils, known as ashless or sulphur/phosphorus oils, are now available for those hydraulic systems employing a pump whose metallurgy is not compatible with ZDDP.

The anti-wear characteristics of hydraulic oils are usually assessed on full-scale pump test rigs by oil companies developing new oils and by pump manufacturers approving particular hydraulic oils.

Figure 7.4   Sectional views of Sperry Vickers Vane Pump rings after bench tests. Left: typical failure with heavy scoring. Right: typical pass showing original mechanical finish

Hydraulic system contamination control

In addition to the essential requirements of the fluid it is also important that some form of contamination control is designed into hydraulic power transmission systems. Contamination in a hydraulic system can arise from different sources, such as:

(1) Residual contamination left in the system after assembly.
(2) Entry with the hydraulic fluid.
(3) Entry with atmospheric air through the reservoir oil breather.
(4) Wear and corrosion particles.
(5) Decomposition of the hydraulic fluid.
(6) Incompatibility of seals and gaskets with the hydraulic fluid.

Figure 7.5   Axial piston pump silver-plated valve-face after pump test,
showing loss of silver plating

Filtration of the hydraulic fluid is the most effective maintenance procedure
that can be used to prevent malfunctioning of the system. This is especially
true in the case of modern numerically controlled machine tools that have
repeatability functions to fulfil. Regular checks of the level of contamination in
such hydraulic systems, and oil and filter changes when indicated, are necessary
to ensure trouble-free operation. The general trend in hydraulic systems is towards finer filtration (in the region of 5 μm or less) as equipment manufacturers
recognise the advantages of clean systems.

Fire-resistant hydraulic fluids (FRHF)

These fluids resist or prevent the spread of flame, either by means of their water
content or by their chemical content. They have different properties from
mineral hydraulic oils, and these must be catered for if a system is to operate
satisfactorily. The chief differences are in density, which may cause pump cavitation in systems designed for mineral oils, and in seal and paint compatibility.
The type of FRHF needed will depend on the application, and cost will enter
into the selection. The main types of FRHF may be summarised as follows.

Oil-in-water emulsions

The oil-in-water emulsion has water as the outer phase with tiny particles of oil dispersed throughout. The low lubricity of water is improved slightly by the addition of a soluble oil to form a homogeneous emulsion. Typical concentration of the soluble oil in water is 5 per cent, hence another name for this type of fluid is 5/95. Soluble cutting oils have been used for many years to blend this category of fluid. Such 5/95 fluids are successfully used in the metal-working industries at pressures up to 27 MPa where the hydraulic equipment has been designed to run on mainly water.

High water-based fluids

Today there are a growing number of 'second generation' purpose-formulated hydraulic fluids containing typically 95 per cent water which are becoming known as HWBF (high water-based fluids). New additive technology has enabled these HWBF to have better anti-wear properties than the first generation cutting oils. HWBF are now being used in hydraulically operated machine tools designed for mineral oils with some modifications at pressures up to 7 MPa.

Water-in-oil (invert) emulsions

Water-in-oil emulsions have a creamy consistency: oil forms the continuous phase, and the water (about 40 per cent) is dispersed as fine droplets (0.5 μm). In general, water-in-oil emulsions can be used in hydraulic systems designed to operate on mineral oils. Systems running on water-in-oil emulsions tend to run at lower temperatures than when mineral oils are used.

Water–glycol solutions

The water–glycol fluids are solutions, not emulsions, since glycols and their additives are truly soluble in water. A typical fluid comprises 40–45 per cent water with the remainder a blend of glycols and a water-soluble thickener.

Synthetic fluids

This category of fluids is now mainly confined to phosphate esters. They give better lubrication than the fluids containing water and are also suitable for operating at higher temperatures. Special sealing and paint materials must be used with them.

Hydraulic fluid users considering introducing FRHF and, in particular, considering the change-over procedures necessary, are strongly recommended to study reference [4].

HYDROSTATIC DRIVES

Hydrostatic drives are available in package-type units and can also be supplied as separate pump and motor units. The latter arrangement has the advantage that the motor can be placed at some distance from the pump exactly where the power take-off is required.

The principle of operation of a typical hydrostatic drive (Dowty Dowmatic) is shown in figure 7.6. This consists of a variable displacement hydraulic pump connected by two high-pressure pipes (E) to a fixed displacement hydraulic motor. The pump is normally coupled to a diesel engine, electric motor or other prime mover, whilst the motor is usually connected to a final drive reduction gearing.

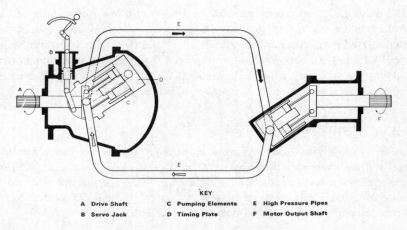

KEY

A	Drive Shaft	C	Pumping Elements	E	High Pressure Pipes
B	Servo Jack	D	Timing Plate	F	Motor Output Shaft

Figure 7.6   Principle of operation of Dowty Dowmatic hydrostatic transmission

The pump is of axial piston tilting head type and the pumping assembly can be moved by a servo jack (B) through an angle of 35° on either side of the neutral position. Altering the tilting head angle will adjust the stroke of the pistons. The resultant variations in delivery rate and direction of oil flow establish the direction of rotation and speed of the motor up to a maximum 1:1 input/output speed ratio condition.

Some of the advantages offered by hydrostatic transmissions are given below:

(1) Exact match of work load and input power.
(2) Smooth quick speed changes from full forward through zero to full reverse.
(3) Control of output speed under all conditions.
(4) Full torque available at creep speeds at low input power.
(5) Output speed virtually constant despite wide load variation.
(6) Dynamic braking.

One of the most important aspects of the hydrostatic drive is its flexibility and it can be used for a wide variety of applications such as fork lift trucks, road rollers, cranes, winches, machine tools and crawler tractors.

Lubrication of hydrostatic drives

The fluids used in hydrostatic transmissions range from engine oils to automatic transmission fluids and, most important, include industrial hydraulic oils. Generally before a hydrostatic transmission manufacturer gives an approval for a particular fluid he carries out bench tests on it. Therefore it is important to check with the transmission manufacturer which fluids are approved and the correct viscosity required for a given application.

HYDRODYNAMIC DRIVES

There are two common hydrodynamic (sometimes called hydrokinetic) drives — the fluid flywheel and the torque converter.

The fluid flywheel or coupling consists of two elements, the impeller and the turbine, which are filled with a medium, almost invariably mineral oil. The impeller imparts kinetic energy into the medium which passes into the turbine and releases its energy in rotating the turbine. At no time is there any increase in the torque transferred but there is some loss in speed. The fluid flywheel also acts as a shock absorber in the line.

The torque converter works on a similar principle but in this case if the turbine is slowed down relative to the impeller there is an increase in torque transmitted. In other words the torque converter behaves as a smooth stepless gearbox and automatically increases the gear ratio as the turbine is slowed down relative to the impeller. The principles of the torque converter are outside the scope of this chapter.

While the fluid flywheel is filled with oil, the torque converter generally has a circulatory oil system and the elements are filled with oil initially by an oil pump.

For suddenly applied loads the torque converter has obvious advantages. As the load is applied the torque converter instantly and automatically alters the gear ratio to compensate for the increased load.

There are obvious advantages using a torque converter in the transmission when one considers the action of say a shovel loader or fork lift truck. The vehicle accelerates towards the load. As the fork or shovel gets under the load the resistance violently increases and, as the truck slows down, the gear ratio is automatically altered in the torque converter and the forward tractive effort is increased accordingly. No time is wasted or momentum lost while the driver has to select lower gears as the truck slows.

There are similar advantages using torque converters in bucket excavators. The bucket needs to be manipulated at high speed to the work-place. As soon as the bucket starts digging a high force is required, and when the bucket is full it has to be withdrawn rapidly to transfer the load. These changes in operation are automatically achieved with the torque converter. The torque converter also acts as a most effective shock absorber in the drive line, isolating the engine from digging shocks.

The torque converter is relatively inefficient, especially when it is under high torque multiplication conditions. However, this mechanical inefficiency is insignificant compared with the increase in machine working efficiency. A torque converter can make drastic reductions in working time in earth moving and material handling. The diesel fuel does not carry a high tax like that for road vehicles, and fuel costs are a small item in total working costs. It pays to have an inefficient transmission and use more fuel if the amount of work done by the machine is increased.

POWERSHIFT GEARBOXES

The torque multiplication in a normal torque converter is around 3:1 maximum, although there are various ingenious modifications to improve on this. This gear ratio is generally insufficient to provide an adequate speed range when working light combined with maximum effort for the heavier duties.

It is usual to provide a wider range of gear ratios by adding an additional gearbox behind the torque converter. This can be a manually selected stepped gearbox but it is generally of the 'powershift' clutch type. These gearboxes enable gears to be engaged with a flick of a switch. One gear is engaged as the other is released, and there is no passing through neutral or stopping if reverse gear is required. The effect of this on work output in 'shuttle' or 'to and fro' operations is obvious.

These powershift gearboxes generally consist of constant-mesh gears which are engaged by hydraulically operated multi-plate clutches. The clutches are closed or engaged by oil pressure behind an end piston controlled by a small valve in the driver's compartment.

These clutch packs consist of alternate bronze and steel plates. The whole clutch is immersed in oil, which is used to cool the plates and prevent wear on the engaging faces.

The powershift gearbox is generally attached to the torque converter and the same oil is used for the hydraulic medium in the converter as used for the lubricant in the gearbox and the hydraulic pressure system in the gearbox operation.

PROPERTIES OF AUTOMATIC TRANSMISSION FLUIDS

At this stage it is opportune to consider the tasks the oil has to perform.

Specific gravity

The oil is the energy-transfer medium in the torque converter. This is the kinetic energy given by the product of the mass of oil and the (velocity of oil)^2. Ideally the specific gravity of the oil should be as high as possible to increase the mass and thus the energy transfer. Unfortunately, as the specific gravity of mineral oil increases so does the viscosity, and specific gravity is of minor importance compared with the low-viscosity and low-temperature properties that govern the low-temperature operation.

Viscosity

As the oil is circulated in the torque converter continuously, and as it transfers energy, it must be of sufficiently low viscosity to flow with a minimum of drag. Viscous drag absorbs power, causes heat and lowers the mechanical efficiency of the converter.

Low-temperature properties

The oil must possess good cold-temperature properties — low pour point and low viscosity — so that it will circulate at the lowest start-up temperatures. At the highest operating temperatures the oil must retain sufficient viscosity to lubricate the bearings in the converter.

It may be necessary for machines in Canada or Scandinavia to operate in mines and then move into arctic temperatures. Alternatively, fork lift trucks can operate outside in tropical climates and then move into deep freeze cold stores. The oil must cope with these extremes of ambient temperatures.

Compatibility

The oil must have no adverse effect on oil seal or O-ring material.

Oxidation and thermal stability

The low mechanical efficiency of torque converters has already been mentioned. At stall conditions when the load is just too great for the transmission and the turbine is held stationary the oil is absorbing the full power of the engine; the transmission is in effect a hydraulic brake. The energy absorption is very high under these conditions and the oil heats rapidly. Oil coolers are provided and often oil temperature warning or shut-down devices are included to guard against these conditions.

In a motor car the automatic transmission normally never approaches stall conditions. In a shovel loader this condition can be approached once in every work cycle. The oil must stand up to these intense heating conditions without serious degradation.

Foaming

The oil must have good anti-foaming properties. A mixture of air and oil has a very low specific gravity and will not transfer sufficient kinetic energy for the converter to work.

Use in gearbox

If the same oil from the torque converter is also used in the gearbox additional properties may be needed. The gearbox requirements are as follows:

Load-carrying properties

The oil must provide adequate lubrication of the bearings and gears under all temperature conditions. This may entail additional EP or load-carrying properties to improve rolling gear performance.

Viscosity

The oil must retain sufficient viscosity at the maximum operating temperatures to work the gearbox operating hydraulic system. If there is insufficient oil pressure at the clutch operation cylinders, the clutch will slip and then fail rapidly.

Dispersancy

The oil must not cause glazing of the clutch plates by degradation products and so cause a lowering of the coefficient of friction. This property is connected with the high-temperature oxidation stability of the oil. It is usual to employ dispersant additives to minimise glazing or coating of the clutch plates.

Compatibility

The oil must not adversely affect the oil seals and O-ring materials.

Viscosity Index

Many automatic gearboxes have completely automatic gear changing. To give acceptable road performance the speeds at which gear changes are made must be varied according to the method of driving. If the vehicle is driven hard, gear changing is effected at higher speeds than if the vehicle is driven gently. In simple terms the speed at which the gear change occurs is generally governed by the oil pressure generated by the gearbox operating pump. This pressure can be affected by the viscosity of the oil − falling off when the oil is hot and the viscosity low. To minimise these temperature effects it is usual to treat oils for these applications with potent viscosity-improver additives to give exceptionally high viscosity indices.

Table 7.2 summarises the additives used in automatic transmission fluids.

Automatic transmission fluid specifications

Ford and General Motors (GM) dominate current specifications for passenger car automatic transmission fluids, which are also widely quoted in commercial applications. In the past there has been a distinct difference between Ford and GM in terms of fluid frictional characteristics, although other properties like viscosity and oxidation stability discussed above are similar.

Traditionally, to obtain satisfactory gearshift quality, Ford have required a high coefficient of friction approaching clutch lock-up. GM required − and still require − a decrease in friction as clutch sliding speeds approach zero.

The friction decrease effect is achieved by friction modifiers. It is illustrated in figure 7.7. This figure shows that Ford now also require such an oil: 'Type CJ' specifications SQM-2C9010-A and ESP-M2C 138-CJ. This is needed for their larger C-6 transmissions but they intend modifying the rest of their trans-

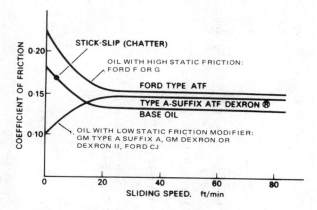

Figure 7.7   Coefficient of friction/sliding speed

Table 7.2 Various additives used in automatic transmission fluids

Additive type	Type of compounds used	Reason for use
Oxidation inhibitors	Zinc dithiophosphate, aromatic amines, sulphurised phenols, etc.	To retard oxidative decomposition of the oil which can result in varnish, sludge and corrosion
Dispersants	Varies	To maintain cleanliness by keeping oil-insoluble material in suspension
Metal deactivators	Zinc dithiophosphate, etc.	To passivate catalytic metal surfaces to inhibit oxidation
Viscosity Index improvers	Various polymers	To lower the rate of change of viscosity with temperature
Anti-wear agents	Zinc dithiophosphate and organic sulphur and chlorine compounds, sulphurised fats, certain amines	To reduce friction, to prevent scuffing and seizure, and to reduce wear
Rust inhibitors	Metal sulphonate, fatty acids and amines	To prevent rusting of ferrous parts during storage and to protect against acidic moisture accumulated during cold operation
Corrosion inhibitors	Zinc dithiophosphate, metal phenolates, basic metal sulphonates	To prevent attack of corrosive oil contaminants on bearings
Foam inhibitors	Silicone polymers	To prevent the formation of stable foam
Seal swellers		To swell seals slightly, reducing leakage
Friction modifiers	Organic fatty acids and amides, lard oil	To reduce the static coefficient of friction

mission systems so that 'Type CJ' fluids can be used across the range. Ford type F or G fluid (specifications SQM-2C9007-AA and ESE-M2C 33G) will still, however, be required for other automatic gearboxes such as Borg Warner.

The movement by Ford from F to G and GM from Type A Suffix A, through Dexron to Dexron II generally represents improvements in oil life and oxidation stability.

There is another well known fluid developed for the General Motors industrial Allison transmission. This is the C type transmission now amended to Type C3.

The Allison transmission consists of a torque converter followed by an epicyclic gear train, the gears being selected by multi-plate clutches operated hydraulically. This transmission can be supplied for two modes of operation. In 'on road' operation fully automatic gear changing can be introduced, similar to private cars. The road speed at which the gears are changed is controlled by oil pressure. In these transmissions (AT, HT, MT and V series) high VI oils — such as Dexron or Dexron II fluids — are preferred to obtain consistent results.

Type C3 fluids are, however, an acceptable alternative for AT, HT, MT and V series transmissions provided the temperature recommendations mentioned below are observed.

All other Allison transmissions are used in the semi-automatic powershift condition and the lower VI C3-type fluids are exclusively specified. The specially designed torque converter/powershift oils are recommended for all temperatures above -23 °C. To assist oil rationalisation, Allison also include in their list of approved C3 fluids diesel engine oils of SAE 10W, 15W-40 and 30 grade viscosity. These may be used for ambient temperatures above -23 °C, above -16 °C and above 0 °C respectively.

The C3 specification does not control the static and dynamic frictional requirements of the oil. In fact most C3 type fluids have high static frictional values. In heavy earth-moving equipment, of course, smooth gear changes are not essential — durability is more important. Under road conditions when fully automatic operation may be used, change quality and control is more important. The operation is generally not so severe and the low static friction, high VI Dexron fluids are preferred.

Many different makes of automatic/torque converter systems are now in use. For example there are the Ford and GM systems just mentioned, Borg Warner, Clark, Caterpillar, J.I. Case, Twin Disc, Voith and ZF. Lubricant recommendations vary between manufacturers and they also vary for different designs from each manufacturer. Recommendations range from diesel engine oils of appropriate viscosity to the fluids described above, although normally alternatives are given as in the case of the GM Allison automatic system. By careful study of makers' requirements, the opportunity exists for some degree of oil rationalisation in a fleet of vehicles or fork lift trucks of different design or manufacture.

REFERENCES

[1] L. G. Woods, The change of viscosity of oils containing high polymers when subjected to high rates of shear, *Brit. J. Appl. Phys.,* 1 (August 1950).

[2] J. K. Appeldoorn, A simplified viscosity–pressure–temperature equation, *SAE Summer Meeting 1963*, Paper 709A.

[3] T. I. Fowle, Aeration in lubricating oils, *Tribology International*, June 1981.

[4] *Guidelines for the Use of Fire Resistant Fluids in Hydraulic Systems*, RP86H, CETOP, London, 1979.

8 Compressor Lubrication

D. F. G. Hampson *C.Eng., M.I.Mech.E.*
Esso Petroleum Company Limited

A compressor is a power-driven mechanism for raising the pressure of a gas by doing work on it. Air is the most plentiful gas, and it is compressed more extensively than any other. Most of this discussion will therefore deal with the compression of air, though many of the principles involved apply also to the compression of other gases. Separate sections will consider the lubrication of compressors used with other gases including refrigerants.

Two considerations that exert a significant influence on the lubrication of air compressors are heat and the presence of water. The magnitude of their effects depends largely on the degree of compression – the pressure ratio. This is the ratio of the absolute discharge pressure (P_2) to the absolute suction pressure (P_1) and it is expressed as P_2/P_1.

When any gas is compressed its temperature tends to increase and most compressors are equipped with cooling devices of one sort or another to moderate the temperature rise. Whilst intercooling (cooling between compression stages) and aftercooling improves the efficiency and lowers the temperatures of compression, they are also responsible for the precipitation of water within the system, the source of water being the water vapour found in atmospheric air. Cooling is therefore limited to maintain temperatures sufficiently low for good lubrication and mechanical operation, but not so low as to cause excessive condensation.

LUBRICATION OF AIR COMPRESSORS

Reciprocating

The lubrication requirements of reciprocating compressors (figure 8.1) can be divided into two parts, the requirements of the cylinders and those of the bearings. Generally, in the small single-acting compressors the crankcase forms a reservoir for the lubricating oil which lubricates both bearings and cylinders. Lubrication of the cylinders is usually carried out by oil splash, excess oil draining by gravity

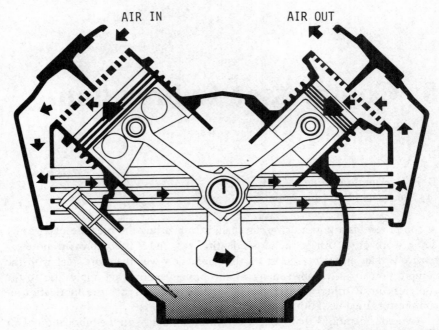

Figure 8.1   Two-stage reciprocating compressor (courtesy of Atlas Copco Ltd)

to the crankcase. The service conditions are not radically different from those
encountered in the automotive engine crankcase. Such single-stage machines tend
to run at temperatures high enough to vaporise moisture in the air being compressed and so avoid contamination of the crankcase oil with water.

As the initial charge of oil is in continual use, even with make-up, the oil
must be renewed at the compressor manufacturer's recommended frequency
(500 − 1000 hours of operation). The main degradation of the oil is by oxidation,
which can be detected by an increase in viscosity, acidity and acrid smell.
Allowing an oil to stay in use above its recommended operational time would
result in thermal breakdown of the oil and possible breakdown of the compressor
as a result of deposits forming on the valves.

Lubricating oils for small reciprocating compressors must have good oxidation
stability and are therefore usually blended from predominantly paraffinic base
oils. This type of compressor is often used in a portable form because of its small
size, and is driven by an internal combustion engine. In many cases the same
lubricant as used in the engine, usually a detergent-dispersant type oil, is used in
the compressor.

In large two-stage reciprocating air compressors the cylinder is separated from
the crankcase by a cross-head and cylinder stuffing box. There is no direct way
for crankcase oil to reach the cylinder as it does in a single-acting compressor, nor

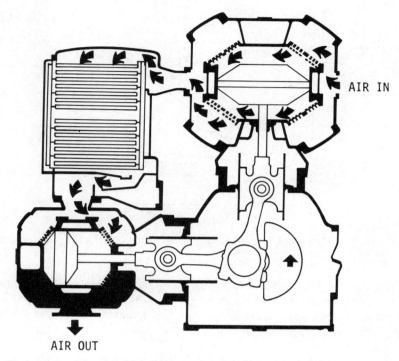

AIR IN

AIR OUT

Figure 8.2   Two-stage reciprocating compressor with cylinder oil feed
(courtesy of Atlas Copco Ltd)

can excess cylinder oil drain back to the crankcase. Two independent lubricating systems are therefore necessary, one for the main bearings crank pin and cross-head assembly, the other for the cylinder (figure 8.2). It is however usual to use the same lubricant in both systems.

Lubricating oil is fed direct to the cylinder walls at one or more points by some form of mechanical force-feed lubricator. By this means precisely metered quantities of once-through lubricant are continually replenished on the cylinder walls. The drive for the force-feed lubricator is taken from some moving part of the compressor so that the lubricator can deliver oil only while the compressor is in operation.

Generally, the viscosity of the cylinder lubricant should be increased with increasing cylinder pressure[1]. However, there is evidence to suggest that the higher the viscosity grade of cylinder lubricant used, the more deposits are formed in the valve chamber and discharge lines. In practice the oil of minimum viscosity necessary to provide adequate lubrication should be used, which is that rec-ommended by the compressor manufacturer. As a proportion of the cylinder feed oil is carried out with the discharged air, the feed rate must be sufficient to pro-vide an adequate film of oil on the cylinder walls.

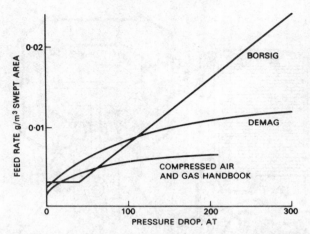

Figure 8.3 Cylinder lubricant feed rates for reciprocating gas compressors

A wide divergence exists between the feed rates recommended by a number of compressor manufacturers, as shown in figure 8.3, taken from reference [2]. It is usual to provide a slight excess feed rate during running-in and then gradually reduce this until on inspection there is no evidence of oil accumulation in valve chambers.

Many cylinder lubrication systems are equipped with sight feed lubricators by which the flow of oil to the cylinders in drop form can be observed. It is important to carry out a regular check, by actual measurement of the amount of oil being fed to the compressor cylinders.

As a small amount of oil is in contact with a large quantity of air in cylinder lubrication, the main requirement is a high oxidation stability. If the oil is not stable, oxidation can produce successively oil-soluble oxidation products, insoluble sludgy and tacky deposits which particularly affect the operation of the valves, and finally carbon deposits. In the early seventies, straight cut narrow distillation range naphthenic oils were recommended as giving the best overall lubrication of reciprocating compressors[3]. However, they did not eliminate the build-up of carbon deposits, especially in large reciprocating compressors with separate cylinder lubrication.

In Europe, concern about the safety of air compressor lubrication has led to a German standard DIN 51506 being accepted by most compressor manufacturers as being the performance criterion of oils used in reciprocating compressors. This standard specifies a maximum level of carbon formation of different viscosity grades after they have been subjected to oxidation or ageing tests (DIN 51352-2). The most severe level of performance in DIN 51506 is designated VD-L and relates to oils for use in compressors with air discharge temperatures up to 220 °C. Well-refined straight cut paraffinic oils containing high-temperature oxidation inhibitors appear to give the best performance against this specification. Field experience of oils meeting DIN 50506 is still being compared with actual compressor operation.

Rotary compressors

Sliding vane

This type of compressor can be either single-stage or two-stage, the latter usually being used for higher air capacities. Sliding vane rotary compressors are manufactured in two types, one being water-cooled, the oil serving only to lubricate and seal. In this type the oil is fed mechanically to rotor and bearings in small

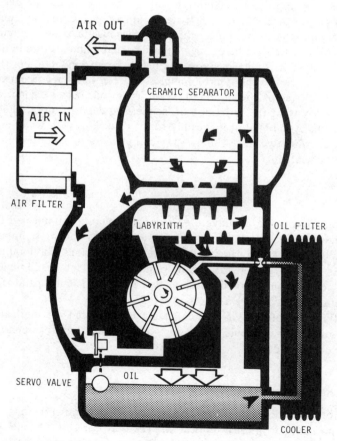

Figure 8.4   Operating principle of oil-flooded rotary vane compressor (courtesy of The Hydrovane Compressor Co.)

metered quantities that are eventually discharged from the system with the out-going air. The other type of sliding vane rotary compressor is oil-cooled as well as lubricated. Oil is flood-injected into the intake air, recovered, and recirculated (figure 8.4).

Both types of sliding vane compression can be considered to have similar lubri-cation requirements, which are met by paraffinic oils of turbine oil quality having good demulsibility. In this type of compressor the oil must withstand the severe oxidising environment in the compression chamber. Further, with the oil-cooled sliding vane compressor oil carry-over should be low so that large amounts of oil are not discharged with air. A low-volatile oil should therefore be used for this type of compressor, and paraffinic oils are less volatile than naphthenic oils of the same viscosity. Paraffinic oils can also be made more resistant to oxidation than naphthenic oils by the use of additives.

There is one important exception to the use of paraffinic-type oils in sliding vane compressors, that is for single-stage compressors in mobile applications. Such machines tend to run hotter than two-stage machines and the temperature of the compressed air is usually high enough to maintain moisture in the vapour state and remove it from the compressor before it can condense. For this reason many manufacturers of this type of compressor prefer oils that contain dispersant-detergent additives of the motor oil type. These oils have excellent oxidation stability and prevent the formation of lacquer and deposits on the sliding vanes. However, this type of lubricant readily forms emulsions with water and should not be used with two-stage sliding vane compressors or compressors that run only intermittently.

Screw compressors

Screw compressors are manufactured in an oil-free and oil-injected form. The single-stage oil-injected type of screw compressor is the most important for compressing air and many are used in portable applications. Oil flooding permits the rotors to run together, has a cooling and sealing effect and eliminates metal-to-metal contact. The oil used is recirculated and an efficient separator keeps the oil content of the discharge air to about 30/35 ppm.

Paraffinic oils of turbine oil quality are usually used for lubricating screw compressors but they should have pour points low enough for operation of the compressor out-of-doors.

Lobe or Roots compressors

As the rotors of this type of compressor revolve without touching each other there is no need for internal lubrication. The only parts of the lobe compressor that require lubrication are the shaft bearings and timing gears. Again paraffinic oils of turbine oil quality are usually used.

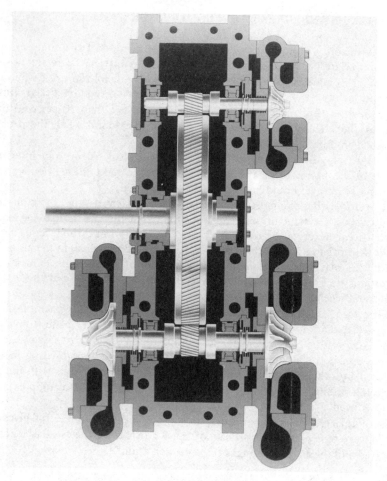

Figure 8.5   Three-stage centrifugal compressor
(courtesy of Joy Manufacturing Co. Ltd)

Kinetic (centrifugal compressors)

Industrial requirements for large volumes of air at low pressures are met by a variety of blowers: kinetic units such as centrifugal (figure 8.5) and axial-flow compressors, or positive-displacement units such as those of the lobe type. With all three types the only motion is rotary — and this without friction between compression-chamber surfaces and without need for sealing. Lubrication therefore applies only to the bearings. The principal demand of the lubricant is long service life, a requirement met by an oil of turbine quality.

EXPLOSIONS AND FIRES IN AIR COMPRESSOR SYSTEMS

For the majority of compressed air systems, lubrication of the compressors by the appropriate type of mineral oil gives satisfactory service. There are, however, critical air compressor systems where there is a fire/explosion risk when using conventional types of lubricant, and these systems do require special consideration. Such critical installations are thought to be less than 10 per cent of the total in the UK and it appears they can only be found when the problems described are found to exist.

Fortunately, fires are more common than explosions but any interrupted service of a compressed air installation that is suspected to have originated from such an occurrence should be thoroughly investigated.

In reciprocating compressors, especially the cross-head type, the compressed air always carries some of the oil into the discharge lines. This oil, being exposed to high temperature, is oxidised and breaks down to form carbonaceous deposits as shown in figure 8.6. If these deposits are allowed to collect in the air delivery lines, fires may occur by spontaneous ignition of the deposits under certain conditions[4]. Compressors are rarely damaged by such fires, but the discharge piping and receiver can be severely damaged.

Fires in oil-flooded rotary air compressors can be caused by lack of oil or by excessive oxidation of the oil. The oil separator filter pads (made from felt) in these compressors can become loaded with oxidised oil and this area of the compressor used to be a common site for fires[5]. Since this type of filter has been replaced by other materials, including porous ceramics, and oils with improved oxidation stability have appeared, the incidence of fires in these compressors has been reduced.

In oil-injected screw compressors, the large quantity of oil injected gives a low air delivery temperature which cannot cause a fire. Nevertheless, flash fires have occurred and the cause is thought to be oil starvation[6].

Figure 8.6   Carbonaceous deposits from compressed air system taken from (on the left) pulsation damper and (on the right) delivery pipework

In some cases, carbon build-up in a reciprocating compressor air system may be due to poor installation design. One large compressor manufacturer recommends designing the pressure system on the output side of the delivery flange of the compressor so that any oil carried out cannot remain for long in the hot zone. The firm also recommends fitting a separate aftercooler to each compressor before the discharge line enters a common service main. This has the effect of good oil transportation and good pulsation dampening with little possibility of oil collecting[3].

Another approach to the problem is that of ICI who have found that the rate of deposit formation in reciprocating compressor systems is negligible if air delivery temperatures are restricted to 140 °C. Delivery temperatures of 140 − 160 °C are accepted provided there is regular inspection of the delivery lines to ensure that deposits are not allowed to build-up progressively.

Currently, the types of mineral-oil lubricant to use in reciprocating compressor systems are paraffinic oils meeting DIN 51506 VD-L quality. Oils meeting this specification are also being used in rotary compressors[7]. However, an oil-ageing test for mineral oil lubricants used in oil-flooded rotary compressors is still to be finalised and accepted by both compressor manufacturers and users.

For the last few years there have been signs that synthetic compressor lubricant use is increasing in both rotary and reciprocating compressors, especially in the USA. These special lubricants are more expensive than conventional mineral-oil based lubricants, but their use may be justified on the grounds of extended service.

Usually, synthetic lubricants are based on di-esters but an alkylated benzene-type lubricant has been marketed in Europe for the last 12 years for use in cross-head type reciprocating compressors. This alkylated benzene lubricant has solved many critical deposit problems and even has a cleansing action on old carbon deposits left in the discharge system from conventional mineral-oil lubricants.

LUBRICATION OF GAS COMPRESSORS

There are many instances in which the nature of the gas under compression makes special lubrication procedures necessary. Rotary compressors of the type in which the bearing lubricant does not come into contact with the gas being compressed are used for large volumes and low-pressure service. For higher pressures reciprocating compressors are used and may require a special lubricant if the gas being compressed is in contact with the lubricant either in the crankcase or cylinder.

In compressing oxygen, chlorine or hydrogen chloride, because of their reactive nature with mineral oils, oil-free compressors are used in which the piston is fitted with filled PTFE or carbon rings to avoid metal-to-metal contact. In these

Table 8.1   Summary of air compressor lubricant requirements[a]

Type of compressor	Lubricant requirements	Recommended type of lubricant
Small reciprocating		
Splash lubricated from crankcase	Good oxidation stability, corrosion resistance, low pour point for mobile applications. Low carbon-forming paraffinic oils	Hydraulic, turbine and motor oils in non-critical applications where frequency of oil changes is not so important. Critical applications oils to DIN 51506. Long life synthetic lubricants.
Large reciprocating		
Crankcase	Good oxidation stability, corrosion resistance, demulsibility	Paraffinic oil to DIN 51506. Note that it is usual to use the same lubricant as for cylinder lubrication. Synthetic lubricants
Cylinders	Excellent thermal stability, low carbon-forming tendency, corrosion resistance	Straight cut (narrow distillation range) paraffinic oils with high temperature oxidation inhibitors to DIN 51506. Synthetic lubricants including volatile alkylated benzene type lubricants
Rotary		
Single-stage sliding vane	Excellent oxidation stability, low pour point, corrosion resistance. Should not be volatile	Paraffinic turbine oil, also HD (detergent) engine oils
Two-stage sliding vane	Excellent oxidation stability, low pour point, corrosion resistance good demulsibility. Should not be volatile	Paraffinic turbine oil type and oils to DIN 51506
Screw	Good oxidation stability, low pour point, corrosion resistance. Should not be volatile	Paraffinic turbine oil type and oils to DIN 51506, also long life synthetic lubricants
Lobe or Roots		
Centrifugal and axial flow	Good oxidation stability, corrosion resistance, good demulsibility	Paraffinic turbine oil type and oils to DIN 51506

[a]   Choice of viscosity grade depends generally on the operating temperatures involved. The higher the temperature the heavier the grade. Consult the compressor manufacturer's operating manual for viscosity recommendations.

cases the crankcases are normally filled with conventional mineral oils because the oil is not in contact with the gas being compressed.

When compressing dry, relatively inert gases such as carbon dioxide, nitrogen, helium, neon and other gases such as air, hydrogen, ammonia and methane, conventional mineral oils of about SAE 30 or 40 grade can be used as cylinder lubricants.

For compressing hydrocarbon gases, such as butane, propane, butadiene and ethylene, an SAE grade 40 or 50 can be used as cylinder lubricant. However, if the gas is wet or at the dew point, hydrocarbon condensate may tend to wash off the lubricant and cause excessive piston and liner wear. Compounded oils with tallow and lard oil have also been used successfully in such applications.

Mixtures of these gases (LPG) in so-called closed-type compressors (that is, where the crankcase is sealed against ingress of air and has an LPG atmosphere) can cause failure of the crankshaft bearings owing to a reduction in viscosity of the crankcase oil. This is a result of the high solubility of the gas in conventional mineral oils. The solution is to use a lubricant of the polyalkalene glycol type which has much less tendency to hydrocarbon gas absorption and consequently retains its working viscosity.

Table 8.1 summarises the lubricant requirements of air compressors.

REFRIGERATION COMPRESSOR LUBRICATION

Refrigeration compressors can be divided into two broad categories, corresponding to industrial and domestic units respectively. Compressor types used are reciprocating and positive-displacement rotary (includes screw). The vast majority of industrial units are comparatively non-critical as regards lubrication, since the lubricant can be changed at regular intervals.

Domestic refrigeration systems usually employ fractional horsepower, hermetically sealed compressor units. These sealed units are expected to run trouble-free for 10 years or more, at higher temperatures and bearing loads than those prevailing in industrial units. They are comparatively critical and require a high-quality lubricant.

Lubricant requirements

The three main criteria which determine the suitability of a refrigerator oil in a refrigerating system are:

(1) its low-temperature characteristics,
(2) its thermal stability,
(3) its lubricating ability.

It would be desirable to specify the physical and chemical properties of a refrigerator oil in terms of standard test methods. Properties such as colour, viscosity and pour point are easily determined by standard test procedures. However, many other properties important to this type of lubricant have not been standardised. A number of non-standardised properties may be assessed by special tests. Although standard tests and special laboratory tests are carried out on refrigerator oils, the performance of a particular oil as a refrigerator lubricant can be fully determined only under working conditions in an actual refrigerator compressor. Compressor manufacturers carry out such tests, especially with domestic units, but they are a lengthy and costly undertaking.

Low-temperature characteristics

Refrigerant 12 is perhaps the most common refrigerant in use today. This refrigerant is completely miscible with mineral oils, but at low temperatures, depending on the oil concerned, wax separates from the mixture. If this occurs in a refrigeration system, its operation may be impaired with blockage of vital components by wax deposits. The tendency for an oil to precipitate wax is defined by its R 12 floc point[8] and the R 12 insoluble content[9].

Figure 8.7 shows the types of wax floc that can be observed in the R 12 floc point test with different refrigerator lubricants. Both tests assess the wax-precipitating tendency of an oil when mixed with Refrigerant 12.

The pour point of a refrigerating oil is also an indication of its low-temperature

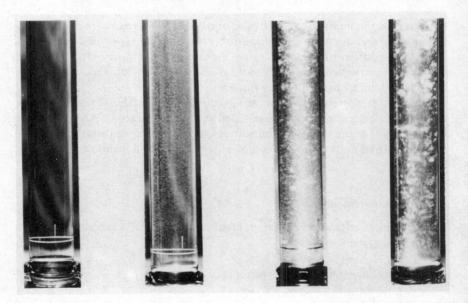

Figure 8.7   Types of wax floc observed in Refrigerant 12 floc point test. From left to right: Refrigerant 12 only; light floc; medium floc; and heavy floc

properties but can be affected (depressed) by small quantities of refrigerant in solution in the oil.

Refrigerant 22 is finding increasing use as it has a lower boiling point than R 12 and gives greater refrigeration capacity for a given size of compressor. However, it is miscible with mineral oils over only a limited temperature range.

Figure 8.8 shows a typical mineral oil/Refrigerant 22 phase diagram. Above the R 22 critical solution temperature, oil and refrigerant are miscible in all proportions. When mixtures are cooled below this temperature they separate into two liquid phases, the actual separation temperature depending on the composition of the mixture. It is thought that the composition of the mixture changes in different parts of the refrigeration system and figure 8.8 shows that the critical solution temperature is lowest at the lower concentration of oil.

In all refrigeration systems some of the oil used to lubricate the compressor is carried over into the rest of the pipework and heat-transfer equipment. In industrial systems efficient oil separators are fitted; these collect the oil that is carried over into the refrigeration systems and so prevent it coating such items as evaporator surfaces with wax (R 12 systems) or congealed oil (R 22 systems), which would reduce the refrigeration efficiency.

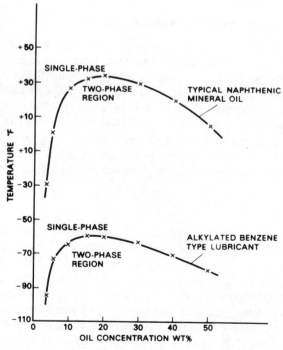

Figure 8.8   Solubility relationship of two types of refrigerator
lubricant and Refrigerant 22

Thermal stability

Refrigerator oils used in both domestic and industrial refrigeration systems require good thermal stability in the presence of the refrigerant being used. Oil degradation is accelerated by exposure to excessively high temperatures, causing a build-up of carbon on such items as valves. Deposits are also thought to originate from the reaction products of various oil and refrigerant fractions, catalysed or initiated by moisture, system metals, etc.

Lubricating ability

Most refrigerator compressors are not excessively loaded and should be lubricated with the lowest viscosity oil that gives efficient lubrication. This is usually decided by the compressor manufacturer and is based on design considerations and service experience. Using the lowest allowable viscosity also means that the low-temperature characteristics are better than those of similar oils of higher viscosity. Compressor manufacturers in Europe do not approve the use of additives in refrigerator oils. This is due to the possibility of the additives separating and giving rise to deposits that could affect critical parts of the system.

Types of compressors used for refrigeration duties

Generally, refrigeration compressors are similar in design to those used for compressing air and can be classified into two main types: reciprocating and rotary compressors. Reciprocating compressors are manufactured in both single and two-stage forms. Rotary compressors can be either single or multi-stage types. Usually the single-stage are the rotary vane type and are often used as a booster, feeding a reciprocating compressor acting as the second stage. Another type of rotary compressor is the turbo or centrifugal compressor which is often of multi-stage construction. In recent years the oil-injected screw compressor has also been used in increasing numbers for large-capacity refrigeration systems.

Choice of lubricant

Naphthenic-type base oils are usually used for refrigerator compressor lubricants; because the wax content of these types of oil is low it is desirable to distil a refrigerator oil within a narrow viscosity range (close cut). The oil is then de-waxed to ensure that wax deposits do not settle out of the lubricant when subjected to its lowest working temperature. Refining a refrigerator oil is a carefully controlled process as it is necessary not only to remove undesirable constituents but also to preserve those natural constituents that impart good thermal stability to the oil.

Many oils with high Refrigerant 22 critical solution temperatures work quite satisfactorily for evaporating down to −40 °C because the system incorporates highly efficient oil separators and the proportion of oil to refrigerant being carried round the system is low. However, others do not, and require special lubricants that have greater miscibility. It is common practice now for such systems to use alkylated benzenes which have excellent low-temperature miscibility with Refrigerant 22, as shown in figure 8.8. This type of lubricant also has better miscibility characteristics with other refrigerants such as R 502 and R 13 B1 which have high critical solution temperatures with conventional naphthenic-based lubricants. There are also many other types of refrigerants used for particular applications and their physical properties are given in the British Standard listed below as reference [10].

REFERENCES

[1] D. Summers-Smith, Tribology Convention, 15–17 May 1968, *Proc. Inst. Mech. Eng.*, 182 (1967–68) part 3N.B.

[2] D. Summers-Smith, *An Introduction to Tribology in Industry*, Machinery Publishing Co., London, 1970.

[3] J. Munck, Safe maintenance of compressor plants, *Engineering*, (January 1967) 71.

[4] D. G. Renowden, Fires and explosions in compressed air units, *J. Brit. Fire Services Assoc.*, 1. No. 4 (October 1971).

[5] E. M. Evans and A. Hughes, Lubrication of air compressors, *Conf. Industrial Reciprocating and Rotary Compressor Design and Operational Problems*, Institute of Mechanical Engineers, London, 1969-70.

[6] P. D. Laing and A. G. Russell, Fires in oil injected screw compressors: their prediction, analysis and prevention, *Inst. Mech. Eng., Convention: Fluid Machinery Failures*, University of Sussex, 1980.

[7] Hans W. Thoenes, Safety aspects for selection and testing of air compressor lubricants, *25th ASLE Annual Meeting, Chicago, May 4-6, 1970*, AM.5E.1.

[8] W. O. Walker and W. R. Rinelli, The separation of wax from oil refrigerant mixtures, *Refrigerating Eng.*, (June 1941) 395.

[9] *BS 2626: 1975, Refrigerator oils*, British Standards Institution, London.

[10] *BS 4580: 1970, Number designation of refrigerants*, British Standards Institution, London.

9 Lubricating Machine Tools

L. H. Haygreen
Esso Petroleum Company Limited

Consistency in dimensional accuracy and surface finish of parts produced on machine tools is important to the efficient operation and life of those parts in service. It is also important to the speed with which they can be assembled to mating parts on modern high-speed production assembly lines.

The speed, accuracy and cost of parts produced is largely dependent on the rigidity of machine tool structures and workpiece set-ups as a whole, and on the accurate motion of the various bearings.

Lubricants play an important role as they lubricate and cool the large number of different types of bearings, gears and slideway systems. Improved lubricants are being developed continually. Many types of boundary anti-wear and extreme pressure lubricants are available to overcome *stick-slip motion* on slideways, to prevent wear and to withstand high contact loadings and surface temperatures without breakdown.

Many modern machine tools also have hydraulic drives and the same fluid can often be used both as a hydraulic medium and a machine lubricant to simplify and to lower the cost of lubrication.

MACHINE TOOL DESIGN

There is a great need to rationalise machine tool lubrication starting at the design stage. Most large factories have a wide range of machine tools bought from all over the world and the total number of lubricants specified by all the makers is usually high. At times makers even specify lubricants that are not readily available outside their immediate area. This can be troublesome and costly to users, particularly during the machine guarantee period.

Table 9.1[1] gives a classification of lubricants for machine tool applications compiled by the International Standards Organization (ISO 3498 −1979(E)), and Table 9.2[2] a rationalised range of lubricants for machine tool applications compiled by the British Standards Institution (BS 5063: 1982). Designers are recommended to select from this range.

Table 9.1 Classification of lubricants for machine tools (reproduced from ISO 3498: 1979 by permission of the British Standards Institution, 2 Park St, London, W1A 2BS, from whom copies can be obtained)

First families according to code letter	Subdivision in categories according to particular application fields	More particular subdivision	Subdivision in categories according to particular properties	Symbol and viscosity grade ISO-L	Detailed application	Remarks
A Total loss systems			Refined mineral oils	AN 68	Total loss general lubrication of lightly loaded parts	
C Gears	Enclosed gears	Moderately loaded gears	Refined mineral oils with good anti-oxidation properties (straight or inhibited)	CB 32 / CB 68 / CB 150	Pressure, bath and oil mist (aerosol) lubrication of enclosed gears and allied bearings of headstocks, feed boxes, carriages, etc., when loads are moderate	CB 32 and CB 68 can also be used for flood-lubricated, mechanically controlled clutches. CB 68 may replace AN 68
		Heavily loaded gears	Refined oils with good anti-oxidation properties (straight or inhibited mineral oils) and with good load-carrying ability	CC 150 / CC 320 / CC 460	Pressure and bath lubrication of enclosed gears of any type (except hypoid gears) and allied bearings when loads are high, provided that the operating temperature is not above 70°C	They can also be used for manual or centralised lubrication of lead and feed screws and highly loaded slideways
F Spindles, bearings and associated clutches		Spindles and bearings	Refined mineral oils with very good anti-oxidation, anti-corrosion and anti-wear properties	FD 2 / FD 5 / FD 10 / FD 22	Pressure, bath and oil mist (aerosol) lubrication of plain or rolling bearings	They can also be used for applications requiring particularly low viscosity oils, such as fine mechanisms, hydraulic or hydro-pneumatic mechanisms, electromagnetic clutches, air-line lubricators and hydrostatic bearings
		Spindles, bearings and associated clutches	Refined mineral oils with very good anti-corrosion and anti-oxidation properties	FC 2 / FC 10 / FC 22	Pressure, bath and oil mist (aerosol) lubrication of plain or rolling bearings and associated clutches	They are required for lubrication of systems including clutches which involve the use of oils that do not contain anti-wear additives
G Slideways			Refined mineral oils with good lubricity and tackiness properties preventing stick-slip	G 32 / G 68 / G 150 / G 220	Lubrication of plain bearing slideways. They should be particularly useful at low traverse speeds to minimise discontinuous or intermittent sliding of the table (stick-slip)	They can be used for the lubrication of all sliding parts such as lead and feed screws, cams, ratchets and lightly loaded worm gears with intermittent service
H Hydraulic systems	Hydraulic systems		Refined mineral oils with very good anti-corrosion and anti-oxidation properties	HL 15 / HL 32 / HL 46 / HL 68	Operation of general hydraulic systems. Lubrication of plain or anti-friction bearings and gears (hypoid types excepted)	They are also suitable for the lubrication of plain or rolling bearings, and all types of gears normally loaded (worm and hypoid gears excepted). HM 32 and HL 32, HM 68 and HL 68 may replace respectively CB 32 and CB 68
			Refined mineral oils with very good anti-corrosion, anti-oxidation and anti-wear properties	HM 15 / HM 32 / HM 46 / HM 68	Operation of general hydraulic systems which include highly loaded components	
			Refined mineral oils with good viscosity/temperature properties	HV 32 / HV 46	Application in computers	In some cases, HV oils may replace HM oils
	Hydraulic and slideways systems		Refined mineral oils of HM type with anti-stick-slip properties	HG 32 / HG 68	Specific application to machines with combined hydraulic and plain bearing slideways lubrication systems where discontinuous or intermittent sliding (stick-slip) at low speed is to be minimised	They can also be used for lubrication of separate slideways, when an oil of this viscosity is required. HG 32 and HG 68 may replace G 32 and G 68 respectively
X Applications requiring grease	Multi-purpose greases		Greases with very good anti-oxidation and anti-corrosion properties	XM 1 / XM 2 / XM 3	Plain rolling bearings, open gears and general greasing of miscellaneous parts	Grease XM 1 is used in centralised systems while greases XM 2 and XM 3 are dispensed preferably by cup or hand gun. The equipment manufacturer should identify the grease used for the initial filling of each item to ensure that the grease subsequently introduced is compatible with it

Table 9.2 Rationalised range of lubricants for machine tools[a] (reproduced from BS 5063 by permission of the British Standards Institution, 2 Park St, London, W1A 2BS, from whom copies can be obtained)

Class	Type of lubricant	Viscosity grade number (BS 4231)[b] for oils, or consistency number for greases	Typical application	Detailed application	Remarks
Lubricating oils					
CB	Highly refined mineral oils (straight or inhibited) with good anti-oxidation performance	32 68	Enclosed gears and general lubrication	Pressure and bath lubrication of enclosed gears and allied bearings of headstocks, feed boxes, carriages, etc. when loads are moderate; gears can be of any type, other than worm gears and hypoid gears	CB 32 and CB 68 may be used for flood-lubricated mechanically controlled clutches. CB 32 and CB 68 may be replaced by HM 32 and HM 68
CC	Highly refined mineral oils with improved load-carrying ability	150 320	Heavily loaded gears and worm gears	Pressure and bath lubrication of enclosed gears of any type, other than hypoid gears, and allied bearings when loads are high	May also be used for manual or centralised lubrication of lead and feed screws
FC	Highly refined mineral oils with superior anti-corrosion and anti-oxidation performance	10 22	Spindles	Pressure and bath lubrication of plain or rolling bearings rotating at high speed	May also be used for applications requiring particularly low viscosity oils, such as fine mechanisms, hydraulic or hydro-pneumatic mechanisms, electro-magnetic clutches, airline lubricators and hydrostatic bearings
G	Refined mineral oils with improved boundary lubrication and tackiness performance, and which prevent stick-slip	68 220	Slideways	Lubrication of all types of machine tool plain bearing slideways; in particular required at low traverse speeds to prevent a discontinuous or intermittent sliding of the table (stick-slip)	May also be used for the lubrication of all sliding parts as lead and feed screws, cams, ratchets, and lightly loaded worm gears with intermittent service: if a lower viscosity is required HG 32 may be used
HM	Highly refined mineral oils with superior anti-corrosion, anti-oxidation and anti-wear performance	32 68	Hydraulic systems	Operation of general hydraulic systems	May also be used for the lubrication of plain or rolling bearings and all types of gears normally loaded (worm and hypoid gears excepted) and airline lubricators. HM 32 and HM 68 may replace CB 32 and CB 68 respectively
HG	Refined mineral oils of HM type with anti-stick-slip properties	32	Combined hydraulic and slideways systems	Specific application for machines with combined hydraulic and plain bearings, and lubrication systems where discontinuous or intermittent sliding (stick-slip) at low speed is to be prevented	May also be used for the lubrication of slideways, when an oil of this viscosity is required
Lubricating greases					
XM	Premium quality multi-purpose greases with superior anti-oxidation and anti-corrosion properties	1 2 3	Plain and rolling bearings and general greasing of miscellaneous parts	XM 1. Centralised systems; XM 2. Dispensed by cup or hand-gun or in centralised systems; XM 3. Normally used in prepacked applications such as electric motor bearings	

[a] It is essential that lubricants are compatible with the materials used in the construction of machine tools and particularly with sealing devices.

[b] The viscosity ranges of the grades mentioned are:

Viscosity grade number	10	22	32	68	150	220	320
Viscosity range (cSt) at 40°C	9.0–11.0	19.8–24.2	28.8–35.2	61.2–74.8	135–165	198–242	288–352

When machine tools are designed with lubrication in mind, it is much easier for users to schedule lubrication of their equipment. The extra cost involved, usually a small percentage of the total cost of machines, is well worthwhile considering that machine life is extended, less time is required for lubrication and unplanned down-time is noticeably reduced.

The following features contribute to the safe and economic lubrication of machine tools:

(a) Plates on machines clearly showing the grades of lubricant, quantities and intervals required. The machine name, type and model should also be clearly visible.

(b) Lubrication either automatic or required at not less than 200 hour intervals, and reservoirs large enough for that period.

(c) Suitable method of lubrication to minimise or prevent metal-to-metal contact at all times; for example, some principal bearings have *hydrostatic lubrication* where oil is supplied under positive pressure, keeping a lubricant film between surfaces even when they are stationary. Although more expensive than conventional bearings they prevent wear during starting and stopping.

(d) Sufficient lubricant capacity to prevent overheating, and thermal distortion of machine tool structures. When necessary, coolers fitted or reservoirs placed outside machines.

(e) Reservoirs placed so that lubricant checking and replenishing are possible without machine shut-down or hazard.

(f) Visual means of checking oil levels in reservoirs, visible to oilers when they fill them.

(g) Lubrication points in one central position, accessible without removing guards or covers, and interfering with machine settings.

(h) Adequate sealing to prevent contamination, and provision for removal of contaminants collected by the lubricant in use (such as fit filters).

(i) Drain points large enough for rapid emptying, and fitted with suitable shut-off valves. They should be accessible, and far enough from the floor to get a container under them.

(j) Provision for collection of lubricant from total loss systems. It should not drain on to the floor.

LUBRICATION OF BEARINGS

Spindle bearings

Improvements in cutting tools allow higher cutting speeds. This means that spindle speeds are increasing. But only when spindle bearings can operate with consistent accuracy at high speed can high-speed machine tools function satisfactorily.

(When referring to spindle speeds, the diameter of the bearing as well as its speed in rev/min must be considered as both of these factors influence lubrication requirements.)

The four main types of spindle bearings are plain journal bearings, rolling element bearings, tilting element bearings and hydrostatic bearings. These bearings — found on wheelheads and workheads of grinding machines, headstocks and tailstocks of lathes, and arbors and quills — must meet very stringent operating and production requirements from immediate start-up to shut-down of the machine tool. To maintain consistent accuracy for a long time, the bearings must have: (1) good radial stiffness for minimum deflection and accurate location of journals under all working conditions; (2) accurate rotation with low frictional resistance to rotation; (3) the ability to operate without wear at acceptable temperatures over a wide range of loads and speeds.

Plain journal bearings

These consist of a rotating member (or journal), which is usually steel, and a stationary member (or sleeve), which is usually a softer metal like bronze or white metal. The two relatively moving surfaces are of different material to minimise friction if insufficient lubricant gets to them, thereby reducing the risk of seizure and damage to a journal.

Lubrication of plain journal bearings

Theoretically, the ideal lubricant for any given application has a viscosity just sufficient to establish full fluid film conditions at operating speed. This prevents solid contact and wear. It also keeps fluid friction and, hence, heat generation and power loss at a minimum. In practice a lubricant with a little higher viscosity than theoretically needed is usually chosen to provide less opportunity for metal-to-metal contact.

Factors affecting choice of lubricant

The speed, width and clearance of bearings are the design features that primarily affect choice of lubricant. A main shaft or crankshaft bearing with average bearing clearance — for example, the mainshaft of a large mechanical press operating at relatively low speed and under stop/start conditions — will at best operate only under partial fluid film conditions and will be lubricated adequately only by a viscous oil around 600 cSt at 40 °C. In some cases grease may be needed.

On the other hand, a high-speed grinding spindle with average bearing clearance may establish full fluid film conditions at operating speeds with an oil viscosity around 20 cSt at 40 °C.

However, a similar grinding spindle with extremely fine bearing clearance —

'zero clearance' — may be lubricated adequately only by a very low viscosity oil, around 2 cSt at 40 °C.

Full fluid film conditions in journal bearings can rarely be maintained during starting and stopping. A bearing that takes a considerable time to reach peak speed and come to rest from speed has to rely on boundary lubrication — when there is little or no oil wedge between surfaces — some of the time. Under these conditions, wear may be high, but it can be eliminated or reduced to a negligible degree by the use of a lubricant with adequate high oiliness or anti-wear properties.

There are many ways of applying lubricant to plain bearings. Oil systems include splash, pressure fed, ring oiled, wick syphon, oil nipple or hole, and mechanical lubricator. For grease lubrication, a centralised mechanical system or a screw-down cup or a nipple may be used.

Modern tendencies in machine tool design are towards centralised systems which feed oil or grease regularly and automatically to the bearing. These systems may be fitted with warning indicators and cut-out devices which stop the machine if there is a fault, such as an inadequate supply of lubricant.

Shock or intermittent load can occur in a machine such as a heavy-duty mechanical press. Oil of higher viscosity than normally used may be required or one of high oiliness, with anti-wear or extreme pressure properties, may have to be used.

Grease can be used to lubricate spindle bearings where bearing clearance is suitable. It has better 'stay-put' properties than oil and helps to seal bearings against contaminants. It is a good lubricant under boundary conditions. However, grease will not flow like oil and may 'channel' at the bearing surface, or it may not provide the necessary cooling circuit in cases of high heat generation.

In summary, the choice of lubricant for plain journal bearings is determined by many factors — primarily speed, clearance, operational conditions, method of lubrication and the possibility of shock or intermittent load.

Rolling element — ball or roller — bearings

In addition to lubricating the rolling elements, the lubricant protects the bearings against rust and corrosion. Grease can help to keep external contaminants out of the bearing.

In general oil is the best lubricant for rolling element bearings. However, grease is particularly suitable for 'sealed-for-life' bearings. Nevertheless, when rolling element bearings are fitted into a unit such as a gearbox lubricated with one lubricant, it is not the bearing but the other parts of the unit, such as gears and clutches, that usually determine lubrication requirements.

When there is a heavy load on roller bearings, oils and greases with extreme pressure additives are needed.

Further information on grease lubrication is given in chapter 11.

Application of lubricants

It is important not to put too much lubricant in a bearing. When grease is used, not more than one-third of the space in a dry bearing should be filled, leaving room for grease displaced by churning from the bearing track. This applies especially in a high-speed operation, when excessive churning causes high temperatures that rapidly degrade the grease and can seize bearings from lack of lubrication or expansion caused by heating.

Before coating the inside of the bearing cage, make sure that the bearing is clean and dry. Grease will not adhere properly to surfaces wet with oil or coated with anti-rust compounds for storage purposes.

One method of applying oil lubricant is the flood or spray centralised lubrication system – spraying the bearings with carefully directed oil jets. This technique, which is good for bearings with speeds over 3000 rev/min (76 mm diameter journal), ensures an adequate supply of lubricant and helps to remove heat.

Another technique is the use of oil mist and fog (figure 9.1). With these systems lubricant is carried to machine parts in the form of small particles – ranging from 0.25 to 1.7 μm – dispersed in compressed air. This allows the bearing to be used to its maximum capacity because it helps to keep it cool. Oil mist lubrication systems are simple and reliable. They can be used to replace flood or spray centralised lubrication systems. Lubricants with viscosities above 216 cSt at 40 °C will not give acceptable fogs, and a heated fogging unit must be used to decrease their viscosity.

The free-fogging tendency of an oil – the tendency of the oil particles to remain in a state of stable dispersion called 'free-fog' – is an important criterion

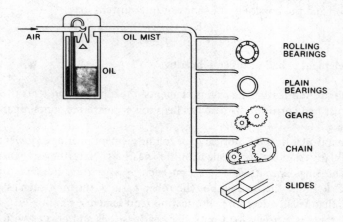

Figure 9.1   Principle of oil mist lubrication

of its suitability for oil fog-systems. Free-fog in the surrounding atmosphere is often a health hazard and always causes oil wastage. Special oils must be used to keep free-fog to a minimum if sealing is not perfect.

At low speeds where the method of oiling is not so critical, oil weir lubrication may be used. With this technique the bearing is immersed in oil. For rotation about the horizontal axis, the bearing should be immersed to a depth not greater than half way up the lowest ball or roller.

Tilting element – rocking shoe bearings

With this bearing the spindle journal is in the centre with each bearing shoe positioned independently around it. The journal and shoes are completely submerged in oil under controlled pressure (see figures 9.2 and 9.3).

When the spindle is rotated, oil enters between the shoes and the spindle, tilting each shoe and forming a wedge-shaped oil film. When the surfaces of spindle and bearing shoes are relatively close together and convergence is in the direction of rotation, the shear forces – forces between the molecules of oil – are high, and pressure is built up. This locks the spindle concentrically within the bearing shoes and creates a high pre-loading of the spindle. Under these conditions the spindle still has perfect freedom of rotation with a very low coefficient of friction. However, because of the high pre-load and because of the ability of the shoes to adjust their wedge angle to accommodate workload, the movement of the spindle under change in workload is only a minute fraction of that which occurs in other types of bearings. A typical viscosity used in this type of bearing is 22 cSt at 40 °C.

Rocking shoes are found mainly in wheelhead spindle bearings on precision grinders and boring machines.

Hydrostatic bearings

With these bearings there is, theoretically, no metal-to-metal contact, even when the spindle is stationary. This eliminates solid friction and other design features prevent excessive float or lift, unacceptable for precision work. Hydrostatic bearings are found in spindle bearings, slideway systems and leadscrew feed arrangements[3].

Lubrication

Hydrostatic lubrication implies that full fluid film lubrication is provided when the associated bearing surfaces are stationary. Liquid or gas can be used as the lubrication medium; oil or air is usually used [4].

Air as a lubrication medium has extremely low frictional characteristics and consequently generates minimal heat. Its advantage is that there is no need to collect and recirculate used air because it can be drawn from and released into

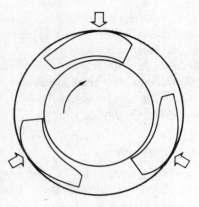

Figure 9.2   Tilting element bearing (clearances are
exaggerated to show the principle)

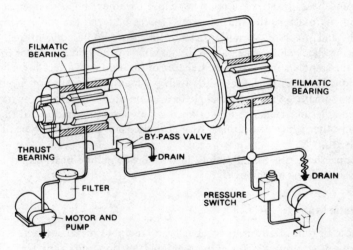

Figure 9.3   Cross-sectional view of the Filmatic bearing
showing the pressure interlock. The spindle cannot be started
until the correct pressure is built up in the bearing chamber

the atmosphere. The disadvantage is that air is highly compressible and therefore
does not provide much bearing stiffness. In addition, really dry and clean air is
essential *and* expensive.

Oil as a lubrication medium has the advantages discussed in chapter 2. A
disadvantage of oil is that it produces a lot of heat as a result of fluid friction at
high speeds. However, frictional heat at high speeds may be minimised by increasing bearing clearances. Another disadvantage is that it is necessary to use a
scavenging system to reclaim and recirculate used oil.

Figure 9.4   Machine tool spindle driven by an electric motor
through a spindle drive gearbox

Oil is usually used at pressures ranging from 3.45 MPa to 20.68 MPa. Oil viscosities range from 2 cSt at 40 °C on very high-speed bearings to 500 cSt at 40 °C for low-speed bearings.

Hydraulic oils are most suitable for these bearings.

Applications

Precision machine tools need adequate stiffness of bearings and minimal heat of friction. Air-lubricated bearings are more suitable for high-speed/low-load applications, oil-lubricated bearings for low-speed/high-load applications.

SPINDLE DRIVES

Machine tool main spindles are usually driven by electric motors, either directly or indirectly (figure 9.4). Spindles are driven indirectly through belts, gears, hydraulic or mechanical stepless speed units, either in combination or independently. Transverse or feed motion of machine parts may be derived from the main spindle drive or an independent power unit.

Figure 9.5   Spur gears and electromagnetic clutches
used in a spindle drive gearbox

Spindle speed and feed gearboxes

Geared or belt drives are most frequently used on machine tools and provide a
stepped speed range. However, stepless (variable) speed drives may gain promi-
nence because they provide smooth and automatic regulation of spindle speed
(see figure 9.5). This helps to maintain constant cutting speed at the surface of
the workpiece, thereby reducing production time and maintaining regular surface
quality. Lubrication of variable-speed drives is critical.

Most machine tool spindle speed and feed gearboxes are comprised of spur,
helical and worm gears. Spur, helical, bevel and worm gears all have one lubrication
characteristic — the resistance to motion between mating teeth. This is partially
derived from sliding friction, which cannot be easily alleviated by a continuous
fluid film.

In spur, helical and bevel gears the direction of sliding is at right angles to the
line of contact. This causes formation of partial fluid film lubrication. Straight

mineral oils are satisfactory under normal load conditions, and it is best to use the lowest viscosity oil capable of carrying the load with a reasonable safety factor.

In some speed and feed gearboxes, in addition to gears, the cams, chains, various bearings, and worms and clutches must also be lubricated. Friction clutches are the greatest problem.

For a friction clutch the required frictional properties at contact surfaces may be impaired by high-viscosity oils and oils containing additives of the EP or solid type. Usually only low-viscosity lubricants around 30 cSt at 40 °C may be used. Friction clutches operating under frequent start/stop working conditions usually generate considerable heat and can affect the stability of the lubricant. Fine wear debris from clutch plates can contaminate the lubricant, which must be carefully filtered to protect both the clutch and other machine elements.

Overload conditions may be expected in the gearing of machine tools such as heavy-duty mechanical presses. For this application, oils specially compounded with mild extreme pressure properties are used. For exposed gears, special adhesive/cohesive gear lubricants with corrosion-inhibiting additives are available. These lubricants prevent wastage and contamination of surroundings.

Further information on gear lubrication is given in chapter 6.

Mechanical friction drives (infinitely variable speed drives)

These drives have attained a high degree of perfection and have been developed into a number of well established designs[s]. Most designs have rolling elements between driving and driven components, and rotary motion is transmitted from one metallic part to another by friction generated through point or line contact. The power transmitted depends on the magnitude of frictional drag and the contact pressure. Large power transmission requires high frictional drag and contact pressure.

Oil is used in these drives to cool, and reduce the wear of, the friction drive assembly as well as to lubricate the main bearings and other parts. However, it must not alter the friction characteristics at the drive assembly and cause slip, which would result in unacceptable variations in the output speed. Therefore, the type and viscosity of the oil must be carefully selected; a typical viscosity is 20 cSt at 40 °C.

Traverse arrangements

When the spindle is driven mechanically instead of hydraulically, machine tool traverse arrangements for long-distance slideways movement are usually a plain leadscrew and nut or a ballscrew in combination with worm and wheel or spur gearing. For short distance movement, a cam and follower in combination with spur gearing is usually used.

The accuracy of linear motion and positioning of machine parts is largely dependent on the accuracy with which traverse arrangements work. Friction and wear on the thread of a nut or rolling element, or on a leadscrew, causes irregular feed of the tool at the surface of the workpiece.

Worm gears

The motion of worm gears is primarily sliding, which nearly coincides with the direction of the line of contact. This motion makes lubrication difficult because it causes oil to be scraped off as the teeth move into mesh. To minimise the problem, heavier oils than those used for spur gears are employed. The presence of oiliness and adhesive additives is good for retaining oil film between mating teeth.

Leadscrew and nut

There are three main types of leadscrew and nut arrangements on machine tools: plain thread, hydrostatic thread and rolling element (recirculating ball). In leadscrew arrangements the leadscrew usually rotates while the nut travels over it. However, on some machines, the opposite applies.

Plain thread type

The nut is usually made from bearing metal, minimising friction under boundary conditions. The predominant action between the leadscrew and nut is sliding, and stick-slip motion can occur in the same way as in ordinary machine slideways. Because the plain thread types work at very low speeds, they tend to squeeze out lubricants and consequently work only under boundary conditions. Under these conditions a lubricant with extreme pressure additives and tackiness agents is most suitable.

Hydrostatic thread type

See hydrostatic bearings.

Rolling element type

Rolling element leadscrews are usually called ballscrews. They give low friction and great stiffness in the traverse, and they can give less backlash than plain thread types. They usually operate at relatively slow speeds and can be lubricated with straight mineral oil or grease in the same way as other ball bearings under normal loads. The nut may be packed with grease. However, leadscrews are usually near slideways, and the same lubricant may conveniently be used for both.

Figure 9.6 Plain bearing slideway and plain thread leadscrew. Wipers attached to the carriage push debris off the slideways

SLIDEWAYS

On machine tools various machine parts, such as work tables, carriages, headstocks, wheelheads and workheads, are required to move on slideways during machining of workpieces. The friction on the slideways greatly determines the accuracy of linear motion and positioning of machine parts, which finally determines the accuracy of workpieces.

There are three main types of slideway arrangements: plain types, rolling element type and hydrostatic type.

Plain bearing slideways

Accurate linear motion and positioning of machine parts is not easily achieved with plain bearing slideways (figure 9.6). No known pair of metals and/or other materials, such as plastic laminates, can achieve accurate linear motion and positioning through the wide range of machine tool working conditions without lubrication. This applies even when low-friction solid sliders are used[6]. Where combinations of load and speed are extreme it is most difficult to obtain accurate motion and positioning.

The sliding motion on machine tool slideways usually makes continuous full fluid film lubrication difficult and, at the best, only partial fluid film conditions are possible. Particularly if the loads on the surfaces are great and the speed of movement is slow, the motion of the sliding surfaces may become non-uniform or jerky. This is stick-slip motion.

The stick-slip problem is very complex. Basically, stick-slip motion would not occur if there were no elasticity in a slideway system. However, elasticity is inherent in materials of propelling mechanisms, and the slider itself. The many factors affecting stick-slip include the rigidity of the propelling mechanism, the damping forces, and the relation between friction and velocity for the slide materials and lubricant involved[7].

There are two prominent friction test equipments for evaluation of lubricants under stick-slip conditions. They are the Cincinnati stick-slip apparatus which is used mainly in North America and the UK, and the Tannert–Wieland indicator used in Western Europe[8,9].

One way to overcome stick-slip motion is to use a higher-viscosity lubricant. When a liquid lubricant is used under light-load/high-speed conditions, hydrodynamic pressure resulting from fluid friction and convergence of surfaces at the slideway interface may cause excessive float of a machine part. Excessive float may be overcome by using a lower-viscosity lubricant.

The problems of slip-stick motion and excessive float are closely related. If a higher-viscosity lubricant were used to overcome stick-slip motion, excessive table float could occur, and stick-slip motion could occur if a lower-viscosity lubricant were used to overcome excessive float. To avoid those problems, a lubricant film must be used that is almost undetectable because it is so thin. Lubricating oils with special properties are used, but they should be carefully chosen so that the viscosity is suitable for the working conditions.

Slideway systems need proper grooves on the bearing surfaces to ensure easy and even flow and distribution of a lubricant over the whole bearing area.

All leading edges at the grooves should be bevelled or rounded to promote a lubricant wedge and to prevent scraping of oil from a fixed surface.

All debris like swarf, and even water, should be kept off the slideways to ensure efficient functioning.

Rolling element slideways

In this type of arrangement, balls or rollers are set between slideway surfaces, thereby achieving the low-friction and accuracy characteristics of rolling element bearing arrangements. However, because it is expensive to manufacture and assemble precision hardened tracks and rolling elements, the rolling element slideway is restricted in use. In addition, it is difficult to maintain accuracy where traverse motion is required over a long distance.

For lubrication see the section above on 'Rolling element – ball or roller – bearings'.

Hydrostatic slideways

Figure 9.7 shows the essentials of hydrostatic slideway operation. Oil is fed under pressure to all bearing surfaces so that there is no metal-to-metal contact. The only friction therefore is the internal fluid friction of the lubricant, and stick-slip is entirely eliminated. Hydrostatic slideways also make positioning of tables more accurately predictable. Although this method of lubrication is more expensive initially than oil wedge lubrication, less power is needed for the drive and savings can be made on this. The accuracy and predictability obtained can also give savings in service.

HYDRAULIC SYSTEMS

Hydraulic systems are used to move tables and rams, to brake, clamp or lock machine parts and workpieces, for copying or profiling units, and for spindle drives[10]. However, the modern trend in machine tool design is to replace hydraulic drives by electric drives.

Because the system is a potential source of heat, the hydraulic oil reservoir and pump unit are sometimes placed outside the main structure of the machine to

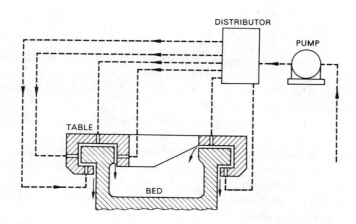

Figure 9.7   Principles of hydrostatic slideway operation

avoid thermal distortion of machine tool parts (which could cause inaccuracies in workpieces) and to make maintenance easier. However, actuating pistons are required in the main structure of a machine tool. Some heat may be generated at these actuators or carried into the machine from the pump unit via the hydraulic fluid.

The following are the main areas of wear and sources of heat in hydraulic systems.

Viscosity

Continuous working of a fluid, despite its low frictional characteristics, results in a rise in temperature as a result of fluid friction. For this reason, it is best to use low-viscosity lubricants when possible.

Internal leakage

Hydraulic systems have a considerable number of moving parts: actuating pistons, control valves, pumps, etc. For maximum efficiency and sensitivity of the system, minimum clearance between moving surfaces is essential to prevent fluid flow — 'leakage' from a high-pressure to a low-pressure area of an actuator. Minute leakage is necessary to lubricate the moving surfaces, thereby preventing solid contact and friction, heat, wear, seizure or inaccurate functioning of the system. However, excessive leakage causes high fluid friction, temperature rise, inaccurate functioning, low pressure and power waste.

Further information on hydraulic systems and fluids, including High Water Base Fluids, is given in chapter 7.

REFERENCES

1   *BS 5063 Rationalised range of lubricants for machine tool applications*, British Standards Institution, London, 1982.
2   *ISO 3498 Lubricants for machine tools – classification*, International Standards Organization, Geneva, 1979.
3   C. M. Stansfield, Hydrostatic bearings for machine tools, Machinery Publishing Co., London, 1970.
4   A. J. Munday, Gas bearings and their application in industry, paper presented at the symposium held in conjunction with the *Industrial Lubrication and Tribology Exhibition, London S.W.1, November 25–28, 1969*.
5   P. Cahn-Spayer, Mechanical variable-speed drives, *Engineers' Digest Surveys Nos. 35 and 36*, April/May 1971.
6   *Developments in the Study of Metal and Plastic Slides*, Production Engineering Research Association of Great Britain, Melton Mowbray.

[7] E. Kadmer and M. Wieland, Stick-slip tests using the Tannert-Wieland indicator, *Mineral oil-Technik*, (1959).

[8] G. J. Wolf, Stick-slip and machine tools, *J. Am. Soc. Lubrication Eng.*, (July 1965).

[9] G. Niemann and K. Ehrlenspiel, Relative influence of various factors on the stick-slip of metals, *J. Am. Soc. Lubrication Eng.*, (March 1964).

[10] C. V. Lizard, Hydraulic variable-speed drives, *Engineers' Digest Survey No. 36*, May 1971.

10 Mobile Plant and Agricultural Equipment Lubrication

A. B. Barr *B.Sc., C.Eng., M.I.Mech.E.*
Esso Petroleum Company Limited
W. Y. Harper *B.Sc., M.I.Mech.E., A.M.I.R.T.E.*

Mobile plant is generally considered to cover the civil engineering contractors' plant — tractors/bulldozers (both rubber-tyred and caterpillar-tracked), scrapers, graders, cranes, cable and hydraulic excavators, air compressors, dump trucks and fork lift trucks.

In basic terms, mobile equipment may be broken down into three units:

(1) Prime mover.
(2) Transmission or drive line. This can consist of gearboxes, torque converters, hydraulic drives, chain drives, wire cables, or compressed air.
(3) At the end of the drive line is the part of the equipment that does the actual work — rubber tyres, caterpillar tracks, digging buckets or paving breakers, etc. This part generally requires no oil lubrication.

The lubrication of the items in (1) and (2) has been discussed in general terms in other chapters. This chapter will look at the particular requirements first of industrial plant and then of agricultural equipment.

ENGINES

Items of mobile plant are virtually all powered by diesel engines. The diesel engine, although more expensive in initial cost, offers considerable fuel economies. Apart from its inherent improved fuel consumption compared with other types of engines, the fuel used is exempt from the considerable taxation levies on all petrol and diesel oil for road transport use.

In chapter 3 the desirable properties of an engine oil are described for various applications. The principal qualities may be defined as:

(1) Flow characteristics — the ability to allow easy starting and to provide sufficient lubricant film thickness and cooling of engine parts at high load by optimum viscometric design.

(2) High-temperature detergency — the ability to keep the piston and rings clean of deposits and operating correctly under full load conditions.

(3) Dispersancy — the ability to maintain the cooler parts of the engine (rocker covers, timing cases, etc.) free from cold sludge and to prevent combustion deposits settling and agglomerating in the crankcase sump by keeping them in suspension in the oil.

(4) Anti-rust properties — the ability of the oil to prevent rust and corrosion while the engine is stationary or running under idling and cold conditions.

The majority of contractors' plant operates under cold-running or inter-mittent-loading conditions. The few conditions of continuous full load occur in scrapers and tractors on heavy bulldozers and chain diggers. In almost all other operations the load is intermittent or cyclic and the average load factor is low. The equipment is almost invariably stored outside with little protection. In many cases oil-storage conditions are poor and contamination with water occurs in bulk storage or barrel storage. Thus the most important lubricant qualities are anti-corrosion and the ability to cope with cool-running conditions.

The obsolete but still quoted MIL-L-2104B or current MIL-L-46152B specifications are designed to produce oil to cope with these basic conditions and the diesel detergency, set by the Caterpillar 1H test, meets most naturally aspirated diesel engine requirements.

Turbo-charged engines are increasingly used in mobile plant and the obsolete Caterpillar Series 3 specification is still quoted for these, although approval was discontinued in 1972. The requirement is now defined by the MIL-L-2104C specification or by API CD classification. These oils give protection under high-load turbo-charged conditions; the low-load factor and cold-running conditions that are typical of plant operation are also met. They are also suitable for most naturally aspirated engines but their relatively high ash contents preclude their use in certain engines.

The General Motors Detroit Diesel engine, which has a two-stroke cycle, is widely used in earth-moving equipment. It is sensitive to engine oil ash content; a maximum of 1.0 per cent mass of sulphated ash is specified by General Motors. Thus, care needs to be taken with these engines if considering the rationalisation on MIL-L-2104C oil throughout a plant fleet.

Multigrade oils can give advantages in cold-starting and cool-running conditions that are typical of plant operations. Most diesel engine builders now approve multigrade oils although some, like Detroit Diesel, place restrictions on their use.

However, most diesel oils give excellent service in utility petrol engines around plant equipment so that the use of multigrade versions gives considerable scope for engine oil rationalisation.

DRIVE LINE

In mobile plant the drive line covers practically every engineering device. The lubricants may be specified by brand names or by specification or description. In a plant depot there will certainly be a variety of oils specified by the various makers of the present items. However it is possible to simplify the requirements in many cases by consultation with the oil supplier. If the various components and their alternative oil requirements are listed, opportunities for rationalisation will appear. Beware against matching branded lubricants as there may be a particular reason why one company has specified a particular lubricant, which is not immediately obvious. The so-called 'equivalent' from another company may not have this particular property.

A brief description of the various components is listed below with their lubricant requirements. Most of these requirements have already been covered in the previous chapters.

Gearboxes

There is a whole range of gearboxes in use in mobile equipment. There are truck-type gearboxes used on dumpers and agricultural tractor conversions. It is usual to call for either an engine oil or an automotive hypoid-type of gear oil for these units. The latter is generally common to the rear axle oil.

Most tractor final drives, crane and excavator hoist, winch and slew gearboxes call for similar oils. Some of the latter boxes may contain bronze worm gears. It is important to check that the hypoid gear oil is suitable for these applications when rationalising a lubricant list.

Some of the larger excavators specify the industrial lead naphthenate type of lubricants for their gearboxes. These oils have high-temperature limitations and are unsuitable for hypoid gearing. While automotive hypoid-type oils may be used in these gearboxes the lead naphthenate-type oils are unsuitable for hypoid gears or hot-running applications.

Some of the larger excavators require very viscous gear oils in their travel gears. The grade will generally stand out as the 'odd oil' but is usually essential for engineering reasons.

Torque converters and powershift transmissions

These are widely used in the more sophisticated equipments. There is a variety of requirements which have been discussed in chapter 7.

Hydraulic drives

Hydraulic drives are now widely used in the construction and materials-handing fields. While the pumps and motors or rams are relatively expensive the problem of connecting the two ends is simply a question of pipe and hosework. The system is thus extremely flexible.

Hydraulic drives can be double-acting; a cable-operated excavator bucket has to drop by gravity, but with a hydraulic machine it is pushed down. Control is simple for the operator. Most small and medium excavators are now hydraulically operated and the work rates are generally considered much superior to the older mechanical models, owing to this double action and ease of control.

In construction equipment and agricultural equipment used in the USA, it is common practice to use engine oils or automatic-transmission oils in hydraulic systems as opposed to the conventional hydraulic oils. These automotive products make excellent hydraulic oils but detergent oils are prone to form stable emulsions if water is present. Many tractor hydraulic systems run hot enough to expel condensed water, but if the system is liable to water contamination the conventional hydraulic oils have better water-shedding properties and may be preferred.

Hydrostatic drives are now widely used on road rollers as they give smooth and accurate control of the rolls. They are also utilised as controls in continuous paving machines used on motorway surfacing. The special oil requirements are covered in chapter 7.

On some hydraulic excavators, constant horsepower control has been fitted to the hydraulic pumps. When the bucket is digging and the operating load is suddenly increased a load-sensitive servo automatically alters the stroke of the pump and enables the hydraulic pressure to be increased. This alters the 'gear ratio' of the system and produces a greater digging effort at the bucket. The result is the same as that achieved by a torque converter in similar applications.

Mobile air compressors

Air compressors are covered in chapter 8. It is often possible on mobile air compressors to use engine oil in the compressor. However, as in hydraulic systems, detergent engine oils under cold conditions can form stable water emulsions and should be avoided.

Airline tools are lubricated by specialised oils, which are fed into the air line or carried in reservoirs on the tool. Some of these lubricants are treated with 'anti-icing' compounds to prevent icing in the tools at points where air expansion occurs, with resultant adiabatic cooling. Icing can be a problem with certain air tools unless these products are used.

Axles

Shovel loaders, wheeled tractors and dumpers all use automotive-type axles. However, because of larger tyres and a lower top-speed requirement the overall gear reduction is much larger than in trucks. This is generally achieved by hub reduction gearing.

Lubricants are generally in the SAE 80W, 90 and 140 classifications for broadly cold, temperate and tropical conditions. The type of lubricant required is almost invariably of the hypoid-additive type with a performance level to the US Military Specifications MIL-L-2105 or MIL-2105C which has a higher anti-scuff rating for the gear teeth.

Some axles may be fitted with 'limited slip differentials'. On an axle fitted with a differential, if adhesion is lost on one side, the wheel spins and traction is lost in the opposite wheel. If the differential is locked the drive is solid across the axle and the non-spinning wheel will continue to drive.

A limited slip differential may be considered as an automatic clutch locking the differential when there is excessive spin. The 'lock' is generally in the nature of a multi-plate or single-plate clutch, which is engaged by the torque reaction generated in the differential when spin occurs.

The normal friction characteristics between two moving oil-immersed plates are shown in figure 10.1.

As the relative velocity between the plates decreases the coefficient of friction increases and reaches a high value (the static friction probably being double the dynamic value). This means that the plates will tend to lock up with a 'snap' as they are engaged or pushed together.

Conversely, as the clutch is released by a reduction in the plate pressure as soon as a breakaway occurs and there is relative motion between the plates, the frictional force falls right away. This can result in stick-slip motion and resulting noise and 'squawk'. This is accentuated because, under marginal slip conditions when one driving wheel is tending to drive and slip, the clutch is feathering or engaging and disengaging continually. The resulting noise from the stick-slip of the clutch can be quite frightening. Axles have been changed in the belief that serious breakage has occurred.

The remedy is to treat the lubricant with friction modifiers so that the static friction is lower than the dynamic friction (see figure 10.1). The process is

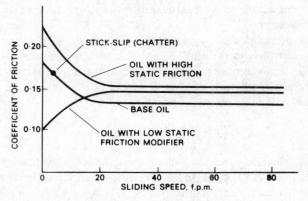

Figure 10.1 Coefficient of friction/sliding speed

similar to that used on automatic transmission fluids to soften the gear shift, see chapter 6.

Road brakes on rubber-tyred construction equipment and agricultural tractors can be adversely affected by mud and water. Very often the hubs are completely immersed in muddy water and slime. This renders the brakes ineffective by lowering the shoe-to-drum friction. When the brakes are released the dirt is pulled round and causes abrasion between shoes and drums. Under these conditions brakes can be worn out in a very short time, or be made ineffective. The obvious course to avoid this is to place the brakes inside the axle casing. This keeps dirt out but prevents effective air cooling of the brake and results in 'brake fade' owing to high temperatures on the brake linings. The next step is to oil-cool the brakes using, if possible, the axle oil.

A problem then arises similar to that in the limited slip differential. Stick-slip motion occurs between brake drum and lining causing noise or 'squawk'. The oil has to be treated with friction modifiers to prevent this noise.

Ford and Massey Ferguson tractors have oil-immersed brakes and the 'wet brake' oils used in their axles are treated to prevent 'squawk'. The oils are also used for the spiral bevel drive and in the hydraulic system and are tailored to give suitable performance in these. John Deere J20A and International Harvester Hytran fluids are also treated to cope with their wet oil-immersed brakes.

There are some engine oils sold in the agricultural field that are treated with friction modifiers and can, in suitable tractors, be used in engine, gearbox, final drive/wet brake/hydraulic systems. None of the fluids discussed above is suitable for use in hypoid drives. There is a limited number of hypoid drives with oil-immersed brakes used on hydraulic excavators. The hypoid-type oils with limited slip differential treatment cope with these applications.

Couplings

In some mobile equipment 'muff-type' couplings are employed between the engine and the driven unit — pump or compressor. They usually consist of two slotted flanges joined by a continuous steel tape threaded through the slots. There are other similar arrangements but the effect is to allow for small relative movements that occur between the flanges as a result of misalignment. The movement tends to be of small amplitude and promotes fretting. It is usual to pack these couplings with grease treated to give EP properties.

Open gears and chains

On some mobile plant there are open gears and chains that require lubrication and protection from the elements. The slewing gears on excavators and the track drive chains are often exposed.

There are no hard and fast rules for lubrication of these items. It depends on the method of application and the conditions under which they are working.

Widely used products for these applications are bitumen-thickened adhesive lubricants. These lubricants are generally applied by brush; some may have to be heated for easy application. They are resistant to water 'washing out'. Their anti-corrosive protection can vary between products. Some bitumastic compounds can be put through suitable automatic lubricators or grease guns and can be used for applications inaccessible to the brush or swab. Solvent 'cut-back' compounds should never be used in extended pipe nipples or lubricators; the solvent evaporates and the compound then chokes the pipe.

Wire cable lubrication

There are no fixed rules about wire cable lubrication; the lubrication depends on the conditions. Bitumastic compounds, oils with tacky additives or light oils may be used according to circumstances.

Bitumastic compounds give good protection for cables working in water but they have difficulty in penetrating between the strands of the cable to provide internal lubrication. For ropes working at high speeds over sheaves, lower-viscosity oils with tackiness additives are preferable. The tackiness additive reduces 'throw off' and water 'wash off'. If cables are dragging through sand, low-viscosity lubricants, which will not pick up sand and carry it into the drums, may be preferable.

On cranes it is a common fault not to lubricate the loose end and drum end of the cable because of inaccessibility. Fixed cables or jibbing cables that are rarely moved still require lubrication and protection from the elements. Even if a cable is not being reeved there is considerable movement between strands as the cable is tensioned and released.

Grease lubrication

Grease lubrication is covered in chapter 11 and only special requirements are mentioned here.

Most slew rings are now of the grease-packed cross-roll or roller-bearing type. They are provided with seals and the whole assembly packed with grease to seal it against moisture and provide lubrication. In some cases, a No. 3 grease or stiff grease is specified as a more effective dirt and water seal. Most makers of these rolls specify EP greases.

For general use, it is customary to use the multi-purpose lithium-based greases, which can stand high temperatures and are water-resistant. It is better to use a higher-quality grease everywhere than stand the risk of a breakdown owing to a lower-quality grease being applied to the one critical application. Few people bother to find out what is in a grease gun!

There is a tendency to specify EP greases for certain highly loaded parts, such as shovel loader pivot pins, and then extend the usage to all applications.

On mobile applications in very cold weather difficulty may be experienced in dispensing No. 3 greases and it is customary to use No. 2 greases in such cases.

In some quarters, there is a tendency to call for specialised and exotic lubricants in certain applications. These specialised lubricants often do extremely good jobs on the applications for which they were designed. The difficulty occurs when the operator has not got the special product to hand, or he has to use a special oil gun to apply it. It is far better to specify a suitable oil, such as say the engine oil that is always available on site, and use it more often, than risk under-lubrication because the special product is not available.

AGRICULTURAL EQUIPMENT LUBRICATION

Because of the rapid growth of mechanisation on the farm, agriculture has led the way in lubricant developments for off-highway equipment. A typical farm lubricant contains anything from 10 to 20 per cent of different additives to cope with the demands of modern engines, transmissions, power take-off drives, hydraulics and braking systems.

Engines

There are two principal components of engine oil performance:

Viscosity, or resistance to flow
Protection against wear, deposits and oil deterioration.

Farm equipment normally requires an SAE (Society of Automotive Engineers) 20 grade or an SAE 30 grade but these single-grade oils need changing to the thinner SAE 10W and 20W grades, respectively, in winter. Failure to make the change may give starting difficulties and could cause engine seizing as a result of oil starvation because the oil is too thick and will not flow to the oil pump when cold.

The need to make seasonal oil changes is eliminated if multigrade oils are used.

Agricultural equipment engine power outputs have gradually increased, so that better protection against wear, deposits and oil deterioration has been re-quired and developed. For example, zinc anti-wear additives help to reduce excessive wear between heavily loaded parts. Also, detergent and dispersant additives help to prevent the build-up of combustion products, sludge and varnish.

The latest turbo-charged engines require oils of high additive content because higher engine loads give greater amounts of carbon and corrosive material that are detrimental to the engine.

The Americal Petroleum Institute (API) and the US Military classifications define oils suitable for modern agricultural equipment engines. For example,

API CD or MIL-L-2104C oils are required for turbo-charged engines while API CC or MIL-L-46152B oils are specified for most naturally aspirated diesel engines and for petrol engines. If API CC oils were used in turbo-charged diesels the long-term durability of the engine would probably suffer. On the other hand, API CD oils may be used in naturally aspirated engines, giving even better protection than required.

Transmission hydraulics

Early farm machinery transmission systems were simple gearboxes and final drives operated by the driver through mechanical linkages. Although transmissions have been vastly improved by hydraulic systems, which will be discussed later, there are still a few applications requiring pure gear oils.

The transmission lubricants cool the gears and provide a friction-reducing and load-carrying film between the working surfaces of gear teeth, chains and sprockets, and for gearshafts with plain or anti-friction bearings.

Several SAE viscosity grades are classified; SAE 75W, 80W, 85W, 90, 140 and 250. Although the numbers suggest that the gear oils are thicker than engine oils, this is not the case. For example, an SAE 80W gear oil has about the same basic viscosity as an SAE 30 engine oil.

SAE 85W−90 and 85W−140 grades are also available. These are referred to as multi-purpose gear oils rather than multigrade mainly because they can be formulated without VI improvers.

Only a few final drive systems require the SAE 140 grade to provide a sufficiently thick oil film to carry the higher loads. SAE 80W or 90 grades are usually sufficient for the average gearbox.

The API and US Military classify and specify gear oils. For example, API GL4 or MIL-L-2105 gear oils have sufficient extreme pressure additives to cope with the loadings of gears up to spiral bevel units. API GL5 or MIL-L-2105C oils have more additives to cope with the lubrication of hypoid gears which have higher loadings and more sliding contact.

Because farm equipment mechanical linkages have been largely replaced by hydraulic systems, pure gear oils have mainly been replaced by hydraulic transmission oils. This is because it is convenient to use the transmission oil as the hydraulic fluid to power clutch operation, power take-off engagement and even external operations like implement lift. Other developments include torque converters, powershift transmission, hydrostatic drives and oil-immersed brakes.

The modern transmission oils are multi-functional and must be capable of acting as hydraulic fluid and have frictional properties suitable for clutch and wet brake operation as well as lubricating gears and bearings.

Wet brakes, operating immersed in oil, require an efficient, noise-free stop by means of friction. This is in opposition to one of the requirements of good lubrication: minimal wear by reducing friction. Special oils with anti-squawk additives are therefore required for wet brake use. Most tractor manufacturers have specifications for these: for example, Massey Ferguson M1135, John Deere J20A, Ford ESEN-M2C-86B and International Harvester Hytran.

These specifications tend to vary, mainly in viscosity requirements. It is therefore impossible to have a single hydraulic transmission fluid that will meet all the parameters of each specification. Most oil companies can supply two or three fluids which cover the range in terms of performance and keep viscosity compromises to a minimum.

Super Tractor Oil Universal (STOU)

The first generation of universal oils for agricultural equipment was named Tractor Oil Universal (TOU). These are still available and are suitable for lubricating up to naturally aspirated diesel engines and also transmissions or hydraulic systems with no oil-immersed brakes or clutches. They are basically API CC type engine oils with additional zinc additives to boost extreme pressure properties for transmission systems.

The later universal oils, called Super Tractor Oil Universal (STOU), can now meet the MIL-L-2104C specification for turbo-charged diesel engines, provide the necessary extreme pressure properties for gear lubrication up to API GL4 level, and have the necessary frictional properties to cope with wet brake and clutch operation with minimum noise and wear.

They are multigrade engine oils suitable for summer and winter use in engines ranging from the smallest petrol one to the most highly rated turbo-charged diesel. They also meet the performance requirements of almost all tractor transmission/ hydraulics and are gradually being accepted by manufacturers. Ford, for example, have introduced a specification ESEN-M2C 159-A for STOUs.

Most major oil companies can supply an STOU grade which has passed performance tests such as MIL-L-2104C for engine oils, MIL-L-2105 for gear oils and various tractor manufacturers' in-house performance tests for transmission/ hydraulic/wet brake oils, although some compromise on the viscosity requirements for the latter may be required.

PRODUCT HANDLING AND STORAGE

Mobile plant lubrication is often performed under uncomfortable and difficult conditions. However, every precaution must be taken to ensure that the lubricants are clean. Bulk storage containers or barrels should be protected from sand and water; barrels should never be stood on their ends — water can enter. All handling equipment, such as measures, must be spotless before filling with oil. Never use the same measures for oil and anti-freeze; anti-freeze in an engine sump turns the engine oil into a black bitumastic-like mess in a very short time. Handling and storage is covered in more detail in chapter 15.

11 Grease Lubrication

E. F. Jones *B.Sc., F.Inst.P.*
Esso Petroleum Company Limited

The essential property that a lubricating grease must possess is the ability to form a film between surfaces so that surface-to-surface contact is prevented [1]. This film may be quite thick as shown in figure 11.1, where grease is being used as a launching lubricant, or it may be very thin as in anti-friction bearings.

If the grease is going to be able to maintain a film of lubricant, certain other properties are required. These can most easily be illustrated by taking an example from the steel industry. Very often grease is the lubricant used in the majority of the sections such as the rolling mill, blast furnace, sinter plant and melting shop. In these areas most of the grease is required for the lubrication of plain or anti-friction bearings. In other parts of the works, gearboxes and linkages will be lubricated by grease.

Wet conditions are caused by high-pressure descaling of the white-hot plate. The atmosphere in the area of the bearings is often heavily contaminated with steam, water, scale and very abrasive particles. In one part of the works the grease will probably be cold and in another part, for example in the region of the hot rolling mills, high temperatures can be expected.

If the grease is going to lubricate successfully it must, therefore, also be able to resist the washing action of water, not soften or decompose unduly over a wide temperature range, retain a good barrier against dirt and prevent corrosion taking place. Most of these conditions require the grease to retain good sealing properties.

Other examples occur where the grease has to remain stable despite vibration or shear, which may soften a grease; for instance the grease used for the axle boxes of high-speed trains is often subjected to vibration. A stage could be reached at which the grease had become so soft that it would be thrown from the bearing and would leak from the axle box. Greases in centralised lubrication systems may also have to be pumped in pipes for long distances under pressure. Continuous pressure may result in oil separation from the grease, hardening of the remaining grease, and possibly blockages of a centralised lubricating system.

Figure 11.1   Applying a thick grease film as a launching lubricant

Apart from the properties already mentioned, the grease must also, therefore, be able to resist vibration, shearing and pressure.

THE STRUCTURE OF GREASE

There have been many definitions of lubricating grease but perhaps the most simple is that used by the ASTM which defines a lubricating grease in the following manner: 'A lubricating grease is a solid or semi-fluid lubricant consisting of a thickening agent in a liquid lubricant. Other ingredients imparting special properties may be present'.

One of the first and still most widely used thickening agents is soap made by the reaction of fatty acids and alkalis. Other products used as thickeners consist of modified clays, specially processed silica and organic compounds. The lubricants that are thickened will also vary from mineral oil or vegetable oils to synthetic liquids such as esters or silicones. Each combination of a particular thickener and a particular lubricant will give the finished lubricating grease particular properties.

By using different alkalis, different soaps can be produced. The main alkalis used for grease-making are lime (calcium hydroxide), sodium hydroxide and lithium hydroxide. These in turn produce calcium, sodium and lithium soaps which give a grease a set of distinctive properties. The fat used to produce the fatty acid will also vary.

As has already been mentioned, a grease must have certain properties, and this means that a grease must have a certain structure. Greases can be likened to a sponge soaked in lubricant. In the case of a soap-thickened grease the 'sponge' would be made of soap and the lubricant of oil. The nature and structure of the 'sponge' will obviously affect the final properties[2].

How then can a definite structure be imparted to a grease? In the case of the premium-quality soap greases this is done by heating the soap in the lubricant until the soap becomes relatively soluble in the oil. Water and glycerine are also produced in this reaction. In the case of soap greases (expect for calcium greases where water is needed to form the structure) the product is dehydrated by driving off the water during the solution stage. Temperatures as high as 250 °C may have to be used. The glycerine remains and helps to promote the solubility of the soap in the lubricant. It also modifies the structure of the finished grease and in many instances imparts a degree of thermal stability to the grease.

After obtaining the soap in solution the next requirement is to cool the hot mixture so that the soap will recrystallise in the desired structure. As the mixture cools so the soap will recrystallise in various forms. It is, therefore, this cooling rate that decides the structure of the grease. Slow cooling may be desirable so that the soap crystallisation, often in the form of very small fibres, can be made to produce a certain structure.

The electron microscope has been of considerable assistance in establishing the structure of greases made by different manufacturing procedures. Small samples are examined using a magnification of 22 000. Certain structures give good service and others give poor service as bearing lubricants. This enables the requirements of the manufacturing procedure to be studied in much greater detail so that the desired structure that performs well in service can be reproduced each time.

So far, discussion has generally concerned soap greases; probably the most widely used non-soap thickeners are the bentonite clays, and preparation of greases from these thickeners is relatively easy.

In general, the consistency or 'thickness' of any grease is determined by the amount of thickener present; in soap greases this can vary from about 5 to 35 per

Figure 11.2   Lubricating the gearboxes of food mixers with a
grease made from non-toxic ingredients for
'sealed for life' operation

cent. The rheological behaviour of a grease will depend on both the thickener
and the lubricant used. Thick greases can be made from low-viscosity lubricants
and vice versa. Other specific properties are given to the grease by the use of
additives.

By selecting a suitable lubricant thickener and manufacturing procedure it
is therefore possible to produce greases with specific properties. This point is
illustrated by figure 11.2 which shows a grease made from specially selected
non-toxic ingredients being used as the lubricant for food mixers.

PROPERTIES IMPARTED BY THE THICKENER AND THE LUBRICANT

The properties that a grease system must possess are, in relation to the ability
to form a seal, water resistance and the ability to resist relatively high tem-

perature without undue softening or melting, properties largely determined by the thickener type. This can be illustrated by considering the soap greases.

Calcium soap greases have good water resistance but because they are stabilised by the presence of water they break down at 100 °C when this water is driven off. This means that they cannot be used for hot conditions. Sodium soap greases, on the other hand, do not melt until the temperature reaches about 150 °C. Sodium soap, however, is soluble in water which means that in the presence of water and oil the soap will form an emulsion. This will eventually lead to washing away of the grease.

Lithium soap greases have the advantage of being water-resistant and also of having fairly high melting points, typically about 180 °C. These greases therefore are often used as multi-purpose greases as they enable several other greases to be replaced by a single soap type[3,4].

Consistency

So far mention has been made only of the characteristics imparted to a grease by using a particular thickener. The consistency of the grease is determined by the amount of thickener present. Because a grease is stiff it does not mean that it has a high melting point. It is also possible to have a soft grease with a high melting point by using a suitable thickener, although it may be semi-fluid at ambient temperatures.

The consistency is measured by penetration. In order to determine the penetration a special pot is filled with grease and brought to a specific temperature. It is placed under a double-angled cone of standard weight so that the tip of the cone is just allowed to touch the surface of the grease. The cone is then allowed to sink freely into the grease under its own weight for 5 s. The test apparatus is shown in figure 11.3. The depth to which the cone sinks in this time is measured in tenths of a millimetre. This is known as the penetration. The softer the grease, the greater will be the penetration.

Because greases have a structure that may change slightly on ageing, it is desirable that the penetration is always carried out with the grease in the same state. This is achieved by giving the grease a standard amount of shearing prior to the penetration being carried out. The grease, in the special pot, is given 60 strokes at a standard rate using a standard perforated disc attached to the top of the pot. The disc passes through the grease with each stroke. This is also shown in figure 11.3. The penetration after carrying out this shearing is known as the worked penetration.

In much the same way as oils are classified in the SAE system, so greases are classified in the NLGI (National Lubricating Grease Institute of America) system by means of penetration. A series of numbers has been given to a series of penetration ranges covering 30 points, with 15 points between each range. The NLGI classification is shown in Table 11.1.

Figure 11.3   Apparatus used to determine the consistency of
a grease by penetration tests. The perforated disc on the left is
used to shear the grease in order to perform
a 'worked penetration' test

Soft products, for example fluid greases, are classified in the 0 grades, medium greases, for example automotive hub greases, in the number 2 range, medium industrial greases, for example high-speed bearing greases, in the number 3 range, and the hard block greases in the number 6 range.

Penetration is only a measure of hardness at one temperature. The way in which the grease softens with temperature is not indicated and no real idea of quality can be obtained purely from the penetration.

Table 11.1   The NLGI classifications of grease consistency

NLGI number	Worked penetration (1/10 mm)
000	445–475
00	400–430
0	355–385
1	310–340
2	265–295
3	220–250
4	175–205
5	130–160
6	85–115

Figure 11.4   Apparatus for 'drop point determination' tests

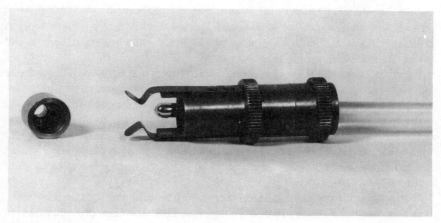

Figure 11.5   Close-up of apparatus from figure 11.4 used to determine the
temperature at which a grease flows under gravity

Drop point

In order to establish what would be the 'melting point' of a grease, a drop point
determination test is carried out. An amount of grease is placed in a small open
cup, fixed on to a thermometer so that the bulb of the thermometer is surrounded
by grease. The cup is heated until a drop falls from the hole in the bottom of the
cup. The apparatus is shown in figures 11.4 and 11.5.

When a soap grease is heated, a stage will be reached when the soap becomes
partly soluble in the oil. At this point the grease will become so soft that it will
flow. A drop therefore falls from the bottom of the open cup. In inorganic thickened greases, for example bentonite, the thickener cannot become soluble in the
lubricant and the grease does not give a drop point.

The drop point is the temperature at which the grease flows under gravity.
This is *not* therefore the maximum temperature at which the grease can be used.
This depends on many factors and there is no hard and fast rule. For example,
the drop point of a simple calcium grease would be about 100 °C but service experience has shown that the maximum temperature for continuous use should
not exceed 60 °C. In the case of lithium hydroxystearate greases the drop point is
about 180 °C but the maximum temperature for continuous operation should not
exceed about 120 °C, although occasional peaks of temperature up to 150 °C can
be permitted.

Resistance to water

The procedure for determining the water resistance of a grease is that a small
bearing is packed with grease and rotated. It is partly protected by a shield, and a

jet of water at a given temperature is directed towards the protected bearing. After a given time the amount of grease that has been washed from the bearing is measured. If a large amount of grease is washed from the bearing the grease will not be suitable for service in very wet conditions.

Water resistance therefore gives some idea as to whether a grease will be washed out when subjected to water wash-out, but it does not give an indication of performance in humid conditions. Sodium soap greases do not give good results in this test but in humid conditions some have been found to give satisfactory performance. One of the important reasons for preventing breakdown by water is that the metal surface may then become corroded. Many sodium-based greases also exhibit good corrosion resistance because they contain specially selected additives.

The general characteristics of the common lubricating greases in relation to their thickeners are summarised in Table 11.2. This table illustrates that a guide can be made to the expected properties of a grease once its thickener is known. It must be stressed, however, that this is only a guide. For example, some sodium soap products are made in a particular way so that drop points of 250 °C can be obtained. Also, it was mentioned earlier that calcium greases are stabilised by the presence of water. If the grease is made so that the water is replaced by a salt or a low molecular weight acid, such as calcium acetate, then the finished grease can be made to yield a drop point of about 250 °C. A similar principle can be applied to lithium greases.

These products are known as complex greases. The load-carrying ability of these greases is also considerably improved without extra additives.

PROPERTIES IMPARTED BY ADDITIVES

The natural properties that greases possess can be improved by the use of additives. While the manufacturing procedure and the ingredients used are often selected to enhance a particular property without the use of additives, other cases arise where additives have to be added.

The selection of an additive will depend on many factors concerning the

Table 11.2   General characteristics of common lubricating greases

	Calcium	Sodium	Lithium	Complex	Bentonite
Appearance	Smooth, buttery	Fibrous	Smooth	Smooth	Smooth
Drop point	95 °C	150 °C	180 °C	250 °C	Not determinable
Water resistance	Good	Poor	Good	Good	Good
Relative cost	Least expensive	⎯⎯⎯⎯⎯⎯⎯⎯⎯⎯⎯⎯⎯⎯⟶			Most expensive

make-up of the grease, for example whether the grease is neutral or slightly alkaline or whether the additive will react with the thickener, etc. Such organic chemicals as certain sulphonates, naphthenates, amines and non-ionic surfactants are commonly used as corrosion inhibitors. In other greases, finely dispersed inorganic chemicals such as sodium nitrite are used.

Other properties that may require improvement are oxidation resistance, load carrying and resistance to scuffing and seizure under boundary-lubricating conditions.

When oils are heated in air, oxidation takes place; the oil becomes dark and develops a characteristic smell. When greases are heated, oxidation takes place in a similar manner and can affect both the lubricant and the thickener, particularly soaps. Although it is readily apparent that oxidation has taken place after heating, oxidation will also occur even at ambient temperatures. In order to slow down oxidation and improve shelf-life anti-oxidants are therefore incorporated in greases. Again the selection of a suitable additive will depend on the lubricant and thickener to be protected.

The fundamental property of a lubricant is preventing surface-to-surface contact. In some conditions very heavy continuous loading may be present. In other circumstances shock loading may occur. Both of these conditions could result in seizure of metal surfaces if the lubricant film is ruptured. In order to prevent this happening the load-carrying properties of a grease often have to be improved. It is convenient to split additives that achieve this into two types, namely load-carrying and anti-wear. This division is, however, only arbitrary.

Load-carrying additives function by means of an active ingredient such as sulphur or chlorine that can be released when a heavy load is applied to the lubricant film, leading to high temperatures where the sliding surfaces come into contact. This ingredient then reacts chemically with the metal surface and prevents seizure of the metal surfaces. The most common additives used in grease for this purpose are sulphurised fatty oils.

Anti-seize additives are usually laminar solids which are incorporated into the grease as such. They are very stable at high temperature and function by forming a film of solid, which has a lubricating ability, between the surfaces. Examples of this type of additive are graphite and molybdenum disulphide. These additives are particularly useful in cases of boundary lubrication and where fretting corrosion can be a problem.

The test methods used for the evaluation of load-carrying or anti-seize properties rely on the building-up of boundary-lubrication conditions. By increasing the load on the rubbing surfaces, a situation can be reached where the film of lubricant between them is broken and welding or scuffing takes place.

The most common piece of equipment for measurement of the load-carrying ability of a grease is the Timken Wear Testing Machine. In this apparatus a rotating test cup is placed in line contact with a test block and the loading between then increased by specified amounts until a scar results on the test block, which shows that the lubricant under test has broken down. The loading at this stage is recorded.

The most common method for the measurement of the anti-seize properties of a grease, on the other hand, uses the Four Ball Test Rig. In this case a ball bearing is rotated in contact with three fixed similar balls so that point contact occurs. A load, which can be measured, is gradually applied until the lubricant film is ruptured and seizure or welding occurs.

Figures 11.6 and 11.7 show scars on Timken test blocks that are just acceptable and not acceptable and balls that have welded after Four Ball Tests.

The testing machines described above are probably more related to development work and quality control than practical conditions and in both cases care should therefore be exercised in relating test results to practice.

A popular misconception is that when molybdenum disulphide is added to a grease, the grease can then be used at much higher temperatures. Molybdenum disulphide does not increase the melting point of a grease and when heated will still run out when it reaches a certain temperature. Should this temperature be reached, then the molybdenum disulphide will probably prevent seizure but if this is likely to happen a higher melting point grease should probably have been used in the first place.

In a comparison of commercially available greases with and without molybdenum disulphide, Risdon and Sargent[5] showed that molybdenum disulphide gave improved strengths to the base grease as reflected by the Falex, load–wear

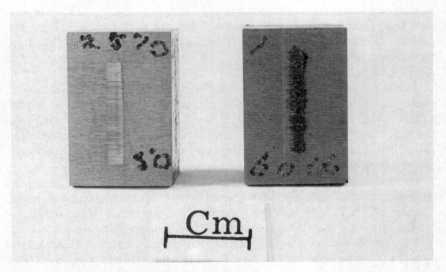

Figure 11.6   Scars on Timken test blocks, just acceptable
and not acceptable respectively

Figure 11.7 Ball bearings 'welded' together after the Four Ball
Test, which is used to measure the anti-seize properties of a grease

index, Four Ball weld load and Four Ball wear scar data. In general, the greatest
benefits were obtained by adding molybdenum disulphide that had the smallest
possible particle size.

RIG TESTING

Because most grease is used in anti-friction bearings much work is carried out in
testing grease in bearings. Obviously the only real way to test a grease for use as
a bearing lubricant, for a particular job, is to run it in the actual conditions. As
this is impractical, various test rigs have been developed that make the running
conditions over a relatively short period fairly severe. There is a very large number
of mechanical test rigs available. Some, like the Institute of Petroleum rig, were
developed largely by representatives of the whole trade, whilst others were
developed by the bearing manufacturers themselves.

The most important factors affecting bearing lubrication are load, speed,
temperature and time. For these rigs therefore a whole variety of conditions for
different bearings is specified according to the view of each manufacturer. For
example, loading conditions vary from a 22.7 kg radial load on the Hoffman
high-speed rigs to 848 kg radial load in the SKF R2F rig. The speed varies from
1500 rev/min in the Ransome and Marles rig to 11 000 rev/min in the Institute
of Petroleum rig. The length of test again varies from 150 hours in the Ransome
and Marles rig to 10 000 hours in the same rig. Because of the length of time
involved in testing, most rig tests are performed for development work. An

Figure 11.8   After 480 hours on the SKF R2F rig, with
a loading of 848 kg at 2500 rev/min, the wear on the
rollers and outer race of this bearing is not acceptable

interesting account of the various rigs used in the development of a multi-purpose grease is that given by Mitchell and Shorten[6].

The various bearing manufacturers also specify the bearings to be used, and the manner in which they should be examined after testing. The condition of the grease after test in also often specified. Figure 11.8 shows a bearing that has completed 480 hours on the SKF R2F rig without external heating, at a loading of 848 kg and 2500 rev/min. The wear that has taken place on the rollers and the outer race can be seen and is not acceptable.

GENERAL PRINCIPLES OF GREASE LUBRICATION

From all the work carried out with test rigs and in service certain principles of grease lubrication for bearings have been established.

When an anti-friction bearing is packed correctly with grease and run, it will eventually reach a steady temperature. It has been found that this temperature largely depends on the amount of grease in the bearing. Performance on the SKF R2F rig illustrates this point. A double-row spherical roller bearing that has been packed in accordance with the manufacturer's instruction will probably rise in temperature from ambient to about 80 °C in one hour when tested in the critical conditions of the SKF R2F machine but will then drop to its steady temperature over the next ten hours. This final temperature will probably be about 20 °C above the ambient temperature. It will then remain at this temperature for a long period. If however the bearing has been overpacked, the initial rise in temperature may also be the temperature recorded during running. The difference in continuous running temperature between a bearing that has been overpacked and one that has been correctly packed may therefore be as high as 40 °C. Hot running is a common cause of bearing failure.

The packing of the bearing is therefore very important. The lowest running temperature consistent with good lubrication is desirable. If the bearing does not have enough grease, on the other hand, the running temperature at first may be quite low but eventually a rapid rise in temperature associated with wear will result. It is usually desirable that the free spaces in the housing and bearing should be filled only to about half of their capacity. There are however one or two exceptions to this rule, where greases perform better when the bearing is fully packed, so no hard and fast rule can apply.

Under the conditions of half-pack, sufficient space is left so that the rollers or balls have a chance to move the main bulk of grease between the balls or rollers into the recesses when the bearing is started immediately after packing. The grease in the area of the balls or rollers is then gradually churned and displaced from the area of movement until an equilibrium is reached. At this point the temperature remains steady – a very thin film of grease is then left to act as lubricant. Scarlet[7] demonstrated, partly by means of dyed grease in the recesses, that very little, if any, movement of grease or oil separating from the grease then comes from the covers to the area of the rollers where lubrication is required. This would indicate that a very thin film of grease has to fulfil the lubrication function without replenishment throughout the period when the bearing is in use.

Other workers have provided evidence to show that lubrication is carried out by the oil that is shed from the grease and in one instance the grease in the bearing was referred to as an 'oil storehouse'.

The true mechanism of lubrication is probably between these two extremes although there are undoubtedly many instances where one theory appears to predominate.

GREASE RELUBRICATION

It is always difficult to assess when a bearing is ready for relubrication as so many factors are involved. The bearing manufacturer should be consulted for advice on the relubrication period and the amount of grease to be used. The SKF General Catalogue for instance gives the following advice.

'The period during which a grease lubricated bearing will function satisfactorily without relubrication is dependent on the bearing type, size, speed, operating temperature and the grease used. The relubrication period is related to the service life of the grease and can be calculated from:

$$tf = k \left(\frac{14 \times 10^6}{n\sqrt{d}} - 4d \right)$$

where

tf = service life of grease or relubrication interval, hours

k = factor depending on bearing type

n = speed, rev/min

d = bearing bore diameter, mm

This equation is applicable to bearings in stationary machines when loading conditions and temperatures are normal and to ball bearings fitted with shields or seals.

The amount of grease required for relubrication is obtained from:

G = 0.00018 DB

where G = weight of grease, ounces

D = bearing outside diameter, mm

B = bearing width, mm'

Other manufacturers give similar information. A chart is available[8] to calculate relubrication periods for Hoffman bearings, based on work carried out in their test department.

In packing bearings all bearing surfaces should be covered with fresh clean grease. In many cases the bearings are not run immediately after packing and the grease is therefore fulfilling the function of a corrosion preventive during storage. If the grease is dirty then damage will result when the bearing is in operation. A survey by Morgan and Wyllie[9] was carried out in which 600 suspect rolling bearings were examined. The most frequent causes of failure were found to be: corrosion; machine and fitting defects; dirt.

If in doubt the bearing manufacturer and lubricant supplier should be consulted.

GREASE OR OIL?

The specific properties that greases possess and how these are used in lubrication have now been covered briefly. The greatest advantage that an oil possesses over grease is that, being a liquid, it has much better heat-transfer properties. This disadvantage on the part of grease is only significant where very high speeds or temperatures exist because modern premium-quality greases have been developed to overcome many of the problems that were traditionally associated with grease. By using grease it is no longer necessary to develop costly sealing systems that would be necessary if oil were to be used. More and more bearings are being lubricated by grease at the present time, which indicates that grease is now accepted as an engineering component in its own right.

REFERENCES

1   E. F. Jones, Lubricating grease, *Tribology*, 1, No. 4 (1968) 209.
2   S. E. Calhoun, Fundamental aspects of grease bleeding, *NLGI Spokesman*, 29 (January 1966) 328.
3   I. D. Campbell and G. L. Harting, A new generation of lithium greases — the lithium complex greases, *NLGI Spokesman*, XL, No. 6 (1976) 193.
4   M. Ehrlich and T. G. Mussilli, The development of lithium complex greases, *NLGI Spokesman*, XLIV, No. 3 (1980).
5   T. J. Risdon and D. J. Sargent, Comparison of commercially available grease with and without molybdenum disulphide, *NLGI Spokesman*, 33 (June 1969) 82.
6   C. H. Mitchell and G. A. Shorten, The development of a multi-purpose lubricating grease, *NLGI Spokesman*, 33 (September 1969) 196.
7   N. A. Scarlett, *Proc. Inst. Mech. Eng.*, 182, Part IIIA (1967-8) 585.
8   Hoffman Manufacturing Co. Ltd, Grease lubrication intervals, *Ind. Lubric. Tribology*, 21 (1969) 304.
9   A. W. Morgan and D. Wyllie, A survey of rolling-bearing failures, *Proc. Inst. Mech. Eng.*, 184 (1969-70) Part 3F.

12 Cutting Oils

A. R. Lindsay *B.Sc., M.Inst.P.*
J. C. D. Russell
Esso Petroleum Company Limited

Although machining operations vary considerably, the basic process of metal removal remains the same. As the tool progresses through the workpiece, metal is compressed until it deforms plastically, that is, it begins to flow and form the chip that slides along the rake face, curls away and eventually breaks off (figure 12.1). A lot of heat is generated – 60 per cent in forming the chip, the remainder resulting from friction between tool, chip and workpiece (figure 12.2). Since the chip and cut surfaces are newly created they are 'chemically' clean and friction is high (see chapter 1). There is a strong tendency for the chip to weld to the tool; in fact small particles do weld together to form a built-up edge. Periodic breaking off of built-up edge mars the cut surface and abrades the tool.

The basic functions of a cutting fluid are:

(1) to remove heat,
(2) to reduce friction,
(3) to carry away swarf.

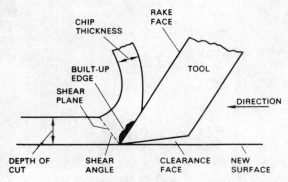

Figure 12.1   Basic cutting action and chip formation

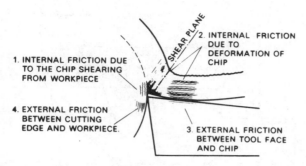

Figure 12.2   Areas of heat generation during machining

A major benefit obtained by the use of a cutting fluid is longer tool life. One way of achieving this is by reducing tool temperature, since a small fall in tool temperature produces a large increase in tool life, for example a fall of 25 °C increases the tool life by 150 per cent[1]. Cutting fluids reduce tool temperature firstly by reducing friction, and hence the amount of heat generated, and secondly by carrying away this heat.

A variety of factors influences the cooling action; most of these are well known. Thus, high specific heat, high thermal conductivity and good heat transfer are necessary for good cooling. Water is twice as effective as mineral oil volume-for-volume, and is one of the most effective coolants known. As will be seen later, watermix fluids are much preferred because of their cooling ability (see figure 12.3) and neat cutting oils are reserved for applications where lubri-

Figure 12.3   Watermix fluids are much preferred because of their cooling ability, as in this milling example

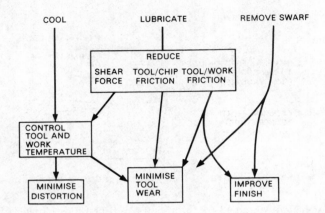

Figure 12.4   Summary of the actions performed by cutting fluids

cation is more important than cooling. Apart from increasing tool life, cooling also minimises distortion of the workpiece and allows it to be easily handled.

In reducing friction between the chip and the rake face of the tool, and between the tool and the workpiece, a cutting fluid must provide boundary lubrication since pressures are such that there is intimate contact between surfaces. The mode of action of oiliness agents and EP additives has already been described (see chapter 1). For minimum tool wear, the appropriate amount of chemical reactivity must be provided that will satisfy the needs of boundary lubrication without causing undue corrosive wear. As the degree of boundary lubrication varies according to machining conditions it is necessary to provide a number of different boundary additive combinations to achieve optimum results — this is one reason for the large number of cutting fluids. The boundary lubrication at the chip/tool interface reduces the welding tendency and hence the rate of wear of the cutting tool. Built-up edge is reduced in size or even eliminated. Thus, lubrication reduces the heat generated, minimises tool wear and improves surface finish.

Summarising then (figure 12.4), cutting fluids facilitate metal removal by lubrication and by cooling. By reducing friction, the amount of heat generated is reduced which, together with efficient cooling, control tool and workpiece temperature thus minimising tool wear. Reduced tool/chip friction also influences tool wear directly and the reduction of built-up edge improves surface finish.

FURTHER IMPORTANT PROPERTIES

A cutting fluid must be pleasant to use. It must not fume, it must not irritate the skin or the eyes, and it must not smell offensively. Much attention has been given to the risk of skin cancer resulting from long exposure to neat cutting oils. To minimise this risk, the major oil companies use only solvent-refined base oils (processed to reduce potential carcinogens) in their cutting oil formulations.

Cutting fluids should not corrode machine tools. This limits the chemical activity of the fluid and watermix fluids must be formulated to inhibit rusting of the machine beds.

Finally, a cutting fluid should be economical. In other words it must remain chemically and physically stable and retain its efficiency throughout a service life that may be as much as 12 months.

CUTTING FLUID TYPES AND THEIR COMPOSITION

Unfortunately, it is impossible to provide the best lubrication and the best cooling simultaneously in the same fluid.

To be an effective coolant a liquid must have a high specific heat, a high latent heat of evaporation and a high thermal conductivity.

	Water	Oil
Specific heat (J/g °C)	4.18	1.88
Latent heat (Abs. J/g)	2257	283
Thermal conductivity (J cm/cm^2 s °C)	0.0063	0.00125

Water is one of the best coolants known; oil is relatively poor. But water has very limited lubricating ability. The result is two distinct classes of cutting fluid:

(1) Neat oils which provide the best lubrication.
(2) Watermix fluids which offer the best cooling.

As a general rule neat oils are used for the more difficult operations and where heavy cuts are taken or where fine finishes are required. These machining conditions generally require cutting speeds of less than 30 m per minute. Above this speed the lubricating ability of a fluid becomes comparatively unimportant relative to cooling because of the greater rate of heat generation; here watermix fluids come into their own. In recent years watermix fluids with enhanced lubricating power have come into use, resulting in their encroachment on the traditional applications of neat oils.

Neat oil composition

The first neat oils to be used were the natural fats, such as lard oil. It was found that a proportion of fatty oil blended into mineral oil gives a more fluid product, which is chemically more stable, and just as good a lubricant. Such oils are still in use today. Some synthetic derivatives of fatty oils are chemically more stable than natural fatty oils and may be used as substitutes.

Where oiliness alone is not enough, sulphur, chlorine and phosphorus additives are included to give EP activity. Flowers of sulphur, dissolved in mineral oil, is the most effective agent for metal cutting but its reactivity is so great that its application is limited to tough ferrous alloys since sulphur in this form will stain yellow metals readily. It should not be used to machine copper or aluminium alloys or in machines that use these metals in their construction. Sulphurised fatty oils are commonly used as a source of sulphur that does not stain yellow metals. Sulphurised sperm oil which showed unique frictional properties for many years has been replaced by technically comparable synthetic sulphurised esters, developed since restrictions of sperm oil were imposed. Although the machining performance of sulphurised fatty oils does not match that of sulphur, a wider application is possible. Chlorinated hydrocarbon is another common EP additive; it is particularly effective for stainless steels and high-temperature alloys. Such additives are non-toxic substitutes for the traditional carbon tetrachloride, which was originally used for high-temperature alloys in the aircraft industry. There are many other sulphur, chlorine and phosphorus additives, some combining more than one element.

It has already been mentioned that it is extremely important to tailor the chemical activity of the fluid to the lubrication requirements of the machining process. An extensive programme of cutting fluid evaluation[2] showed that machining conditions could be compared in terms of *operational severity*, that is, the required levels of lubrication, and that the best tool life at different severities was achieved by different additive combinations:

Degree of severity	Additives
Low	Fatty oils plus small amounts of chlorinated hydrocarbon
Medium	Sulphurised fatty oil
High	Flowers of sulphur plus sulphurised or chlorinated hydrocarbon

The level of severity will be discussed later and is the product of operation (including metal-removal rate), tool material and work material.

For best cooling with a neat oil the viscosity should be as low as possible consistent with freedom from fuming and fire risk.

Watermix fluids

The term *watermix fluids* is applied to concentrates that are dispersed in water for use, and are either *emulsifiable oils* (soluble oils) or *solutions*. Chemical coolants – often called 'synthetics' – may either form true solutions or micro-emulsions. They most nearly resemble solutions and are included in this class.

Emulsifiable oils (sometimes referred to as soluble oils by the sellers and 'suds' by the users) are blends of mineral oil, emulsifier and corrosion inhibitors designed to emulsify spontaneously when added to water to form an oil-in-water emulsion. This emulsion consists of millions of tiny droplets dispersed in water. The emulsifiers are chemicals that encourage the formation of these droplets and discourage them from coalescing.

Under sterile conditions and using distilled water these oil-in-water emulsions would remain stable indefinitely. Water hardness, however, reduces the efficiency of the emulsifiers and tramp oil, fine metal particles in suspension or bacteria, all contribute to the eventual breakdown of the emulsion into oil and water phases.

The concentration of emulsifier determines the size of the oil droplet. General-purpose emulsifiable oils form milky white emulsions, where the droplet is about $1-5$ μm in diameter. By increasing the concentration of the emulsifier, the oil droplet diameter is reduced to below 1 μm, giving a translucent or even transparent micro-emulsion. The mix ratio is defined as the parts of concentrate added to the parts of water. General-purpose soluble oils are used at ratios between 1 to 20 and 1 to 50 for general machining. Transparent emulsions are used at 1 to 60 or above. The greater water content and smaller droplet size increase cooling efficiency and so these products are used where maximum cooling is required, as in finish grinding, for example. Watermix oils usually include a bactericide to discourage the growth of bacteria.

The lubrication afforded by soluble oils, while being better than that of solution fluids, is much inferior to that of neat oils. To overcome this poor lubricating ability, boundary additives like those used in neat oils are incorporated in 'heavy-duty' (or EP) soluble oils. These products go some way to combining good lubrication with good cooling. Their wider use is often inhibited by machine tool design. Older gearboxes are provided with inadequate seals and positioned such that watermix fluids can enter, resulting in rapid wear and sometimes seizure of the bearings.

Modern solution fluids are concentrated solutions of water-soluble corrosion inhibitors which are diluted for use. These are generally metal salts and organic amines. For many years a combination of sodium nitrite and triethanolamine has given a cheap, effective rust-inhibitor system. However, concern over the toxicity of sodium nitrite and the possible formation of nitrosamines with this combination has prompted the development of safer systems. Solutions are excellent coolants, being nearer to water than any other cutting fluid, but they possess very limited lubricating properties. Many machine tools make use of the oiliness of a cutting fluid to keep various moving parts (such as chucks and slideways) lubricated. Neat oils and soluble oils provide this auxiliary lubrication but solutions do not, and the machine tool may become difficult to operate. Another problem associated with solutions is the drying off and resulting sticking of swarf to machine beds. For these reasons solutions are generally confined to high-speed grinding.

Chemical coolants have been devised to overcome the shortcomings of solutions. They use the same types of corrosion inhibitors but also contain other chemicals such as polyglycols, borates and acetates, and some of the conventional EP additives already mentioned, to give oiliness and EP activity. Some of these materials may not be water-soluble and where necessary emulsifiers are added. In these circumstances a transparent micro-emulsion is formed. Some chemical coolants contain mineral oil and so the line of demarcation between heavy-duty (HD) soluble oils and chemical coolants becomes blurred. The main physical difference is that HD soluble oils form milky emulsions whereas chemical coolants make transparent emulsions or solutions. Chemical coolants are the more sophisticated and expensive but, despite their technical advances, chemical coolants are still mainly used for grinding operations; HD soluble oils are preferred for general machining.

In terms of 'relative usage' milky soluble oils are used for 70 per cent of all machining. The most popular neat oils are those used for high-speed automatic machining. The more active neat oils, heavy-duty soluble oils, solutions and chemical coolants are limited to specific applications where their high activity or maximum cooling is strictly necessary.

SELECTING THE RIGHT FLUID

The concept of selecting a fluid with the appropriate degree of lubricating activity and cooling has already been discussed. However, the three influencing factors in making the right choice of fluid are:

(1) Machining operation (including metal-removal rate).
(2) Workpiece material.
(3) Tool material.

All these factors are interrelated in such a way that if one is changed then the influence of others is altered.

All *operations* are variations and extensions of single point turning (figure 12.5). The variables are tool geometry, the number of cutting edges, the area of tool in contact with the workpiece and the rate of metal removal. The rate of metal removal is a function of depth of cut, feed rate and cutting speed. These factors determine the forces on the tool and the amount of heat generated when using a specific work material. Internal operations, such as internal broaching and tapping, are intrinsically more severe in that the tool has a large area in contact with the workpiece, air cooling is prevented and the access of the fluid is

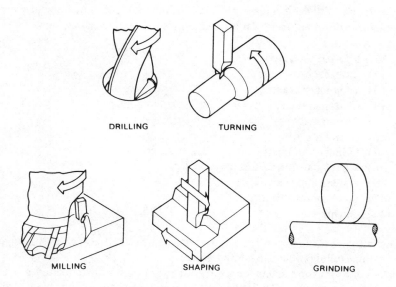

DRILLING   TURNING

MILLING   SHAPING   GRINDING

Figure 12.5   The five basic related machining operations

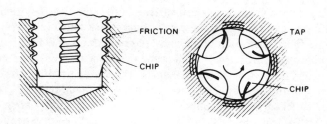

FRICTION

CHIP

TAP

CHIP

Figure 12.6   The tapping operation

limited. Broaches are expensive precision tools, thus making any reduction in wear rate most valuable. Broaching is also critical because both roughing and finishing cuts are carried out with the same tool — hence surface finish is very important.

Tapping is another critical operation. There are many cutting edges in contact with the work, frictional forces are high and wear of cutting edges further increases the area of contact (see figure 12.6). Rubbing friction increases the load and a large amount of heat is generated. With limited access of cutting fluid, heat build-up is great, thus encouraging tap wear. Machining operations are listed below in order of decreasing lubrication requirement. A high metal-removal rate may transform an operation of medium severity into a critical one.

  (1) Internal broaching,
  (2) External broaching,
  (3) Tapping, threading,
  (4) Gear manufacture by hobbing, shaving, milling,
  (5) Reaming,
  (6) Deep hole drilling,
  (7) Drilling,
  (8) Milling and form turning,
  (9) Planing and shaping,
 (10) Single point turning,
 (11) Sawing.

Operations low down on the list are often carried out at high cutting speeds, thus testing cooling rather than lubricating ability.

A notable omission from the above list is grinding and its related processes of honing and lapping. These are the abrasive methods of machining where metal is removed by millions of tiny grits which act as minute cutting tools randomly orientated with respect to the workpiece surface (figure 12.7) such that the cutting is accompanied by much rubbing and a vast amount of heat is generated. A grinding fluid is primarily a coolant and it must also flush away the fine grinding debris which may spoil surface finish and clog the grinding wheel. To maintain grinding efficiency the cutting edges must remain sharp. Periodically, the cutting efficiency of a grinding wheel has to be recovered by dressing with a diamond point. The efficiency of a grinding fluid is often measured by the length of interval between wheel dressings: the better the fluid the longer the period between dressing. Finish grinding, where the metal-removal rate is small and the quality of surface most important, usually employs transparent emulsions, solutions or chemical coolants. Form, or plunge, grinding (the shaping process)

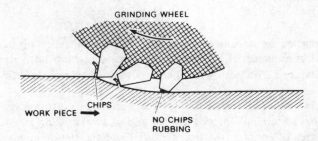

Figure 12.7   The action of randomly orientated grinding grits

is a most difficult operation, where forces on the grinding wheel are very high. A low-viscosity neat oil with good EP activity is employed, avoiding shock quenching which can lead to work cracking.

The other abrasive processes lapping and honing are finishing operations, giving better surface finishes than grinding, and for these operations cooling needs are not so great but lubrication must be good. A typical fluid is a low-viscosity fatty oil blend. Honing can also be used as a metal-removal process, for example in tube manufacture. Here requirements are similar to form grinding and so low viscosity neat oils with EP activity are used.

Some *workpiece materials* machine easily, others are much more difficult. The term most widely used to compare the behaviour of metals in cutting is 'machineability'. This term is difficult to measure because it cannot be directly related to the basic properties of ductility, hardness, chemical composition or microstructure. No method of forecasting machineability from a knowledge of these properties has ever been devised. The reason is that machineability varies according to the method and rate of metal removal. For instance, an increase in cutting speed generates more heat, and so makes the material hotter and softer and easier to cut — unless its ductility is already causing problems, then the increase in temperature will increase the machining difficulty by encouraging the metal to adhere to the tool.

Machineability is most often measured by the efficiency of metal removal, that is, the tool life achieved for a given metal-removal rate. Machineability ratings are expressed as expected tool life for given conditions of workpiece/tool/feed rate/cutting speed/depth of cut.

For general guidance it is possible to list common materials in an approximate order of machineability, ranging from the most difficult (titanium alloys) to the easiest (magnesium).

    (1) Titanium alloys
    (2) Nickel–chrome base alloys
    (3) Inconel
    (4) Nickel
    (5) Stainless steel
    (6) Tool steel
    (7) Monel metal
    (8) High alloy steel
    (9) High carbon steel
  (10) Wrought iron
  (11) Cast iron alloys
  (12) Copper
  (13) Medium carbon steel
  (14) High tensile bronze
  (15) Grey cast iron

(16) Malleable cast iron
(17) Mild steel
(18) Free cutting steel
(19) Silicon aluminium alloys
(20) Aluminium alloys
(21) Unleaded brass
(22) Leaded bronze
(23) Zinc base alloys
(24) Complex wrought aluminium alloys
(25) Leaded free-cutting brass
(26) Magnesium alloys.

Aircraft alloys of titanium, nickel and other metals are developed for high strength above 650 °C. They are all difficult to machine because of their high shear strength; they work harden and they are relatively inert to reaction with cutting fluid additives. Titanium has unusual characteristics; it has a low thermal conductivity and shears in such a way as to produce a large chip/tool interface. Tool temperatures rise so quickly that cutting speeds must be kept extremely low. Titanium also readily adheres to the tool and trace impurities make titanium extremely abrasive.

Stainless steel, like the aircraft alloys, work hardens during cutting and, to minimise this effect, feed rates should be kept as heavy as possible. Extreme ductility can also hinder metal removal. The metal crystals cling together under compressive forces. Plastic flow results and the metal is difficult to shear — the material is referred to as 'luggy' or 'draggy'. The absence of impurities often makes metals extremely ductile. Pure forms of iron, nickel and copper and single-phase alloys, such as commercial bronze, are luggy materials that yield long springy chips.

The composition of a material may be altered to give good machineability. Free machining steels have added amounts of lead, sulphur or phosphorus. The machineability of brass can also be improved by the addition of lead.

There are five main types of *tool material*:

(i) *Carbon steel* and *medium alloy steels* are relatively uncommon because they have been superseded by materials of superior hot hardness and wear resistance. However they can be easily shaped and their principal use is in small form tools operating at low speeds. Watermix fluids are used to keep these tools as cool as possible.

(ii) *High-speed steel (h.s.s.)* tools are still widely used, although they have been superseded by carbides and ceramics for very high cutting speeds. High-speed steel combines good hot hardness and high-wear resistance and is extremely tough. Either neat oils or watermix fluids can be used with h.s.s. tools depending on the requirements of the operation.

(iii) *Cast alloy* tools have a greater hot hardness than h.s.s. tools but less toughness. Typical uses are drills, single point tools and parting off tools. Neat oils are used where possible to protect the tools from heavy shock loading.

(iv) The earliest forms of *carbide* tool were much harder but less tough than h.s.s. Composite carbides of tungsten, titanium, tantalum or niobium improved all-round performance. The most recent significant development is the coated carbide. A tough grade of tungsten carbide is coated with a thin layer of hard, wear-resistant titanium carbide. With such improvements the eventual replacement of h.s.s. in most applications is likely. Carbide tools require the minimum of lubrication and so watermix fluids are generally applied.

(v) *Ceramic* tools have even greater hardness but less toughness than carbides. These are used for very high-speed turning and watermix fluids are used to cool chips and minimise workpiece distortion.

The specific advantages of *diamond* are a low coefficient of friction particularly with soft metals, a low thermal expansion and negligible wear which makes for maximum accuracy of machining. The main application is turning of aluminium pistons and precision bearings.

CUTTING FLUIDS IN ACTION

To get the best results, attention must be given to the correct preparation, application and control of cutting fluids. For long service life, systems must be kept clean, especially with watermix fluids which are most sensitive to contaminants.

Before a machine is filled with fresh cutting fluid, the sump should be thoroughly flushed out to remove all debris and slimes. With watermix fluids, it is best to use a system cleaner, a detergent and/or a solvent, usually containing a biocide, which is added to the emulsion at about 5 per cent and circulated during one eight-hour shift before flushing and refilling with fresh emulsion. Where machine tool sumps are inaccessible, a suction cleaner with different shaped nozzles will tackle this job quickly and effectively.

The correct way to disperse a watermix fluid is to add the concentrate to water (not vice versa) with constant agitation. Semi-automatic mixing can be obtained with a Laycock Fluidmix valve (figure 12.8). Working on the venturi principle, this valve has a variable orifice allowing different mix ratios to be used. Output is around 25 litres per minute of emulsion or solution.

For maximum cooling and lubrication, the cutting fluid should flow copiously over the working area — at low pressure to avoid splashing. Where access of the fluid is restricted high-pressure jets may be applied up the rake and clearance faces of the tool. A common fault is too little fluid and at least 5 litres per minute per cutting edge is recommended.

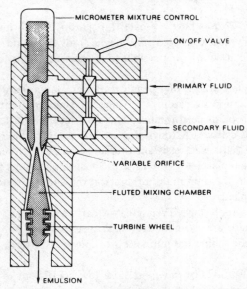

Figure 12.8   The Laycock Fluidmix valve gives semi-automatic mixing
of concentrate and water in variable mix ratios

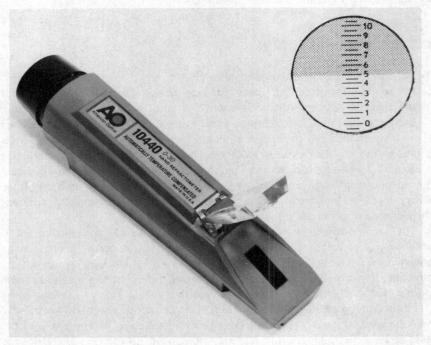

Figure 12.9   The AO refractometer for measuring emulsion strength.
The insert is a view through the eyepiece

Figure 12.10   Refractometer being used in a machine shop

An emulsion or solution may become more concentrated or more dilute during use. Extreme variations in mix ratio may result in fall off in performance. To maintain the correct mix ratio regular checks are necessary. A pocket refractometer is a rapid and effective tool for checking mix ratios of emulsions (figures 12.9 and 12.10). For solutions, titration methods are available which, though less convenient than the refractometer, are still quite simple tests.

The fluid must be kept clean and it is necessary to avoid tramp oil from gearboxes and hydraulic systems, especially with watermix fluids, whose dispersion may become unstable. The presence of metal fines and other debris can catalyse fluid degradation (such as loss of emulsion stability) and cause abrasion of the skin, tool wear and poor surface finish. Filtration will not only extend cutting fluid life, it may radically improve tool life and surface finish. A useful guide to the types of filters to use has been published[3]. Bacteria infection of watermix fluids (infection is relatively rare in neat oils because bacteria need water to live) leads to the familiar Monday morning smell, to instability, to rusting and to staining of machine parts. Most watermix fluids contain bactericides that offer some protection but the best way of avoiding bacterial contamination is to keep systems clean and well aerated.

Finally, to minimise the risk of skin disorders, protective clothing should be worn, splash guards should be provided on the machine tools and frequent washing and application of barrier creams should be encouraged.

REFERENCES

[1] M. E. Merchant, 'Fundamentals of cutting fluid action', *Lubrication Engineering*, August 1960.

[2] J. Beaton, J. M. Tims and R. Tourret, 'Function of metal cutting fluids and their mode of action', *Third Lubrication and Wear Congress, Institute of Mechanical Engineers, London, 1965*, pp. 155-176.

[3] P. J. C. Gough (Editor), *Swarf Handling and Machine Tools*, Hutchinson, London.

13 Metal Rolling Operations

B. A. Cook *B.Sc.*
Esso Petroleum Company Limited

Of the various metal fabricating processes in use today, rolling has become established as the operation that accounts for the highest tonnages. It has proved to be a high-speed process capable of producing materials to close tolerances.

Rolled products come in a variety of shapes and forms. The flat products include plate, sheet, strip and foil, but there are many other sectional shapes, such as rod, wire or tube, that can be fabricated using this technique. The choice of the metal working operation depends a great deal on the properties required in the finished product. Forging and extrusion can confer properties that are unobtainable by rolling and as a consequence these processes have their particular applications.

METAL ROLLING PRINCIPLES

The rolling operation consists of passing metal stock between two counter-rotating rollers. Providing that the gap between them is suitably less than the thickness of the incoming stock, the rolls will grip the material and deliver it at a different area of cross-section. The difference between the entry and delivery cross-sections when expressed as a percentage of the incoming section is known as the 'reduction'. In the case of flat rolled products, a reduction in the thickness of material passing through a mill must be accompanied by an increase in width and length, although many mills are designed to limit the spread (width increase) to a minimum, thereby converting most of the gauge reduction into elongation of the material length. A limited spread is achieved by careful adjustment of the rotation of mill rolls to a faster peripheral speed than that of the incoming stock.

The principles of deformation by rolling are shown in figure 13.1. Metal stock of thickness T_1 enters the mill at a speed of V_1 and is caught by the rolls rotating at the peripheral velocity V. The metal is compressed between the rolls and

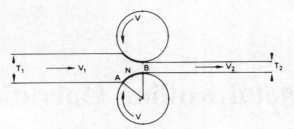

Figure 13.1   The principles of metal rolling

delivered from the mill with a velocity of V_2 and at a thickness of T_2. The velocities of the system are such that $V_2 > V > V_1$ which means that along the contact arc AB the stock speed is increasing and at a point N, its speed will be equal to V and there will be no relative movement between the material and the roll surfaces. Along the arc AN the stock speed is less than the roll speed and this movement of the stock relative to the rolls is known as backward or reverse slip in the mill. Conversely, when the stock moves faster relative to the rolls along arc NB, this is referred to as forward slip. The position of N relative to A and B varies with the percentage reduction as does the length of the arc AB itself. As the reduction increases, the ratio AN/NB increases until a limit is reached where the reduction is the maximum possible. At this limit, N coincides with B; total reverse slippage occurs and the mill is then unable to roll the stock.

MILL LAYOUTS

A typical mill stand consists of two box-shaped mill housings bolted across two bed plates. The separation between the housings is maintained by tiebars. The top plate is sometimes common to both housings. The rolls are located through the central windows of the mill housings as shown in figure 13.2, and the rolls are connected to the drive motors by spindles and gearboxes. Each mill stand contains at least two work rolls and these may be supported by further 'back-up' rolls to assist in distributing the load. The simplest arrangement is the 2-High, but there are complex designs using twenty or more rolls in a single stand. The most common arrangement is the 4-High design and consists of two work rolls with two back-up supporting rolls. A few mill roll configurations are illustrated in figure 13.3.

Frequently, it is necessary to pass stock through several mill stands simultaneously, carrying out successive reductions in each. Such a mill arrangement is known as a continuous tandem mill and the mass flow of stock must be kept constant throughout.

The various operations carried out in metal rolling plants vary greatly in complexity according to the types of products being produced. Figures 13.4 and

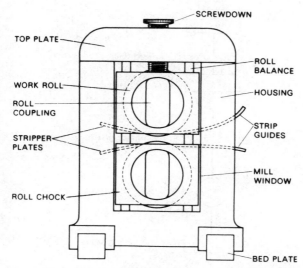

Figure 13.2   Layout of a typical mill stand

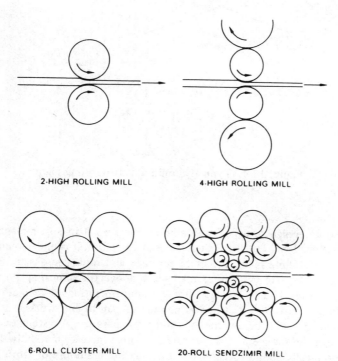

Figure 13.3   Some typical mill roller configurations

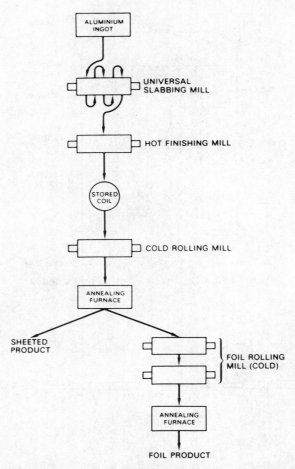

Figure 13.4   Typical aluminium strip rolling plant

13.5 illustrate typical designs for aluminium and steel strip rolling plants respectively. Since steel is a much tougher material to deform it is understandable that the associated rolling plants are more complex and robust in design. Hot metal rolling operations are those that take place above the annealing point of the particular metal. Thus, for steel, hot working operations take place at stock temperatures above 900 °C whereas aluminium can be hot worked at temperatures above 350 °C. Below the annealing temperature, crystalline structures begin to form in the metal and deformation forces increase markedly. It is therefore economically sound to carry out the maximum amount of fabrication that the equipment will allow in the hot state. Cold rolling is generally required to

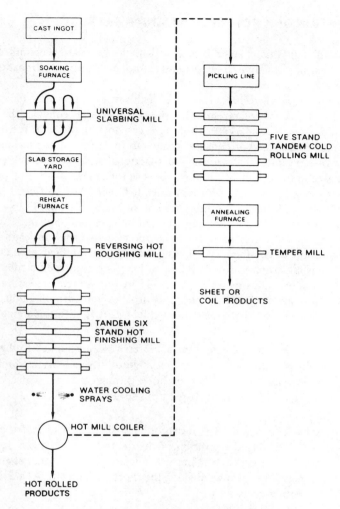

Figure 13.5 Typical steel strip rolling plant

reduce material to gauges below 1.5 mm because the metal becomes too flimsy to work in the hot state. In the cold rolling operation, work hardening takes place, which normally results in a brittle product that would be unacceptable in many applications, particularly when it is required for subsequent forming operations. To remove this brittle hardness the product is annealed; then, if a particular surface hardness is required, the stock is passed through a temper mill which is a very light reduction pass to case-harden the product to the desired specification.

LUBRICATION OF A ROLLING MILL AND ITS EFFECT

The discussion in this chapter is confined to the lubrication of mill rolls only. The roll bearings and hydraulic systems, although requiring lubricants, are discussed in other chapters.

Lubrication of mill rolls is essential if high surface quality products are to be produced without heavy damage to the rolls themselves. Frequently the lubricant has an important secondary function as a coolant for the mill; for, unless the mill rolls are maintained at constant temperature, forces of expansion and contraction will continuously vary the cross-sectional area of the stock (known as the 'shape') and the rolling of products to close tolerances will not be possible.

When a lubricant is applied to mill rolls, it affects the coefficient of friction at the roll bite and, hence, modifies the rate of rolling. The modifying influence will depend on the nature of the lubricant and the amount applied.

In a tandem mill, it is important that the effect of the lubricant should be kept as constant as possible; thus it is necessary to maintain an even application of oil of constant composition. Mill lubrication systems are designed around this principle, to allow adjustments to be made in roll speeds to restore constant mass flow through the tandem train. The visual effects seen on a mill as a result of reducing the friction coefficient through lubricant application are normally:

(1) A decrease in roll force (the compressive stress applied to the stock).
(2) A drop in the motor current required to drive a lubricated mill.
(3) A decrease in mass flow of stock through the mill which can be observed as a decrease in strip output speed.

Referring back to figure 13.1, it can be seen that the effect of the lubricant has been one of shifting the position of the neutral point N towards B for the same roll speed V. There is an increase in reverse slip along the lengthened arc AN and a decrease in forward slip along the shortened arc NB. As a consequence V_1 and V_2 are reduced relative to V.

COLD ROLLING MILL LUBRICATION SYSTEMS

All modern, cold rolling plants are equipped with recirculating lubrication systems and figure 13.6 illustrates the design principles. The lubricant is maintained at constant temperature in a storage tank containing a separating weir. A filter unit is built into the system to remove the metal fines formed in the rolling process. Filtered oil is segregated from the unfiltered returns by means of the weir and the rate of filtration must be higher than the rate of oil delivery to the mill. As a consequence, filtered oil will flow over the weir into the unfiltered stock.

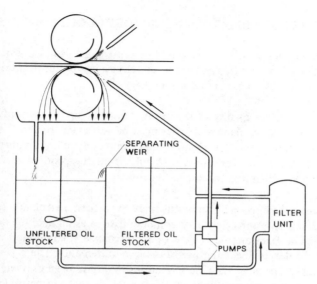

Figure 13.6   A typical flood-lubrication system

Oil is applied through spray nozzles which are spaced regularly across the width of the mill and directed at both upper and lower surfaces of the strip. The system is designed to deliver an excessive flood of lubricant into the roll bite and, in this way, consistency of oil delivery and even application is achieved; the rolling action squeezes away the surplus.

The design of a mill is good from a tribological viewpoint. The rolling action naturally tends to create a wedge of lubricant in the roll bite and the mill behaves like a giant needle bearing. When soft, ductile metals such as yellow metals or aluminium are rolled to a consistent reduction, it is possible to grind the mill rolls such that their mating surfaces deflect parallel under load. Under these circumstances the roller functions at its best and surface variations are minimised. This is the case with the rolling of aluminium foil for example, where the metal is rolled to extremely narrow gauges with high precision and surface quality. On the other hand, when steels of differing hardnesses, widths and gauges are all rolled through the same set of mill rolls, the mill has to be continuously adjusted for each grade and shape of stock being produced. The overall picture is the best compromise possible for the spectrum of products, although slight imperfections in the shape from one grade to another can be expected. Lubrication is also likely to be less than perfect under these conditions and the skill of the mill operator in controlling the mill whilst minimising variations is paramount. Careful scheduling of the grades to be rolled is normally practised on steel mills.

LUBRICANTS FOR COLD ROLLING OF SOFT DUCTILE METALS

Aluminium, copper, zinc and lead would be typical metals in this category. There are a number of factors that influence the choice of the lubricant used in these applications and the main considerations are:

(1) The surface quality of the product must be up to standard.
(2) Lubricant coolant properties must be sufficient for good mill control.
(3) Friction properties must not make mill control unduly difficult.
(4) Lubricant must evaporate cleanly from the product surface without stains when it is annealed after rolling.

These requirements are commonly met by a light petroleum fraction containing a small amount of a fatty compound such as an acid, alcohol or ester. The petroleum fraction must have good resistance to oxidation and thermal cracking in order to meet the non-staining requirements. There is a preference towards highly saturated paraffinic fractions that have been extracted to remove sulphur and other chemically reactive compounds that are naturally present in mineral oils. Paraffinic compounds that boil below 320 °C are generally found to suit this application. It is unwise, however, to use products with low boiling constituents since this will create a fire hazard. Thus the boiling range of the petroleum fraction tends to be cut as narrowly as is economically possible and as high as is practical without causing staining. The fatty additives are normally primary aliphatic acids or alcohols with chain lengths ranging from C_{12} to C_{18}, or their ester derivatives. The addition rate is usually in the region of 1–5 per cent, depending on the compound and the application.

Occasionally other specialised additives are used to suit a particular application. These include oxidation and corrosion inhibitors, polymeric thickeners and anti-wear additives. In the case of yellow metal rolling, copper deactivators may be included. Contamination with tramp oil is a common problem in rolling mills and, in particular, leakages from the mill hydraulic systems. This creates a trend towards hydraulic and roll oils that are compatible without detriment to their relative performance criteria.

The coolant properties of petroleum-based roll oils are found to be adequate in the majority of cases although some of the latest high-speed mills may benefit from the development of new rolling fluids with better thermal-transfer properties because of the increased severity of the operation.

LUBRICANTS FOR COLD STEEL ROLLING

Steel is a much tougher material to deform than most of the common non-ferrous metals. The severity of this operation normally requires the use of lubricant with more efficient coolant properties. To meet this requirement, soluble roll oils have been developed where the oil phase is emulsified into water.

Soluble roll oils contain the same types of ingredients as the neat oils, but with the addition of suitable emulsifiers. In most cases the oil-in-water concentration lies in the range of 1–10 per cent and the emulsion has a milk-like appearance. A recirculation system providing flood lubrication would again be used in this case, with the roll oil stored at elevated temperatures of about 50 °C. The stability of the emulsion is often found to be an influence on product surface quality since it determines the degree of lubrication achieved at the mill bite. Optimum results are usually obtained when the stability is such that the emulsion will partially break down under the pressure and temperature conditions prevailing in the mill bite. This optimum is indicated by maximum load carrying with minimum friction coefficient and the effect is illustrated in figure 13.7. Highly stable emulsions produce high mill loadings with rolling properties tending towards those of water itself. Unstable emulsions produce variable rolling results depending on the amounts of oil that actually get taken into the roll bite and the degree of emulsion separation in the supply system. The formulation of a good soluble oil involves taking account of these factors.

One special problem encountered with soluble roll oils is microbial degradation. The conditions under which emulsions are stored and the abundant supply of food and water present in them make an excellent environment for bacterial growth. Many types of emulsifiers are particularly prone to bacterial attack,

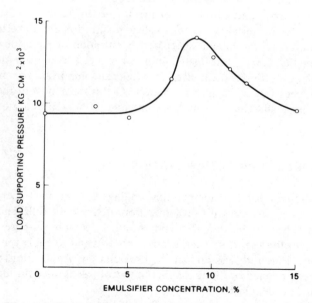

Figure 13.7   Effect of emulsifier content on load-carrying
properties of a soluble cold roll oil

resulting in irreversible damage to the roll oil. Biocides sometimes have to be incorporated into soluble-oil formulations to counteract the problem. This addition can itself create a problem in the disposal of used emulsions since certain biocides cannot be tolerated in effluent processing plants.

LUBRICANTS FOR THE HOT ROLLING OF ALUMINIUM

The annealing temperature of aluminium is sufficiently low for it to be success-fully hot rolled using flood lubrication and a recirculation system. Aluminium is normally hot rolled at temperatures within the range of 350–550 °C, and since emulsions are not rapidly degraded by these rolling conditions, they have proved to be popular in this application also. Modern aluminium hot-rolling emulsions tend to contain synthetic materials with good thermal stability and load-carrying performances. These ingredients have largely overcome one major hot-rolling problem, namely, the pick-up of aluminium particles on the surface of the steel rolls which subsequently become re-rolled into the surfaces of later batches of product. Surface defects that occur at the hot-rolling stage tend to persist right through the subsequent cold-rolling stages and give rise to final products which are sub-standard. Another surface defect that can occur at the hot-rolling stage is the formation of white or brown stains. Problems of this nature are best avoided by physically removing as much surplus emulsion as possible after rolling.

In spite of the use of biocides and filtration, emulsions will slowly deteriorate from bacterial activity and a build-up of metal fines in suspension. The rate of deterioration is thus determined by the quality of filtration and the effectiveness and concentration of the biocide. In practice, filtration of emulsions becomes difficult when the sizes of the metal fines are reduced to the sub-micron pro-portions of the oil droplets themselves. It is therefore important to remove fines effectively immediately after their formation. At this stage most of the particles are relatively large and filterable, but become micro-fine with repeated grinding from the rolling process.

HOT STEEL MILL LUBRICATION SYSTEMS

The hot mills used for high-temperature rolling of steel or copper were not originally intended for use with oil lubrication of the work rolls, and ancillary equipment has had to be designed and installed for this purpose. The mill layout precludes the use of oil recirculation systems and to apply the lubricants as a flood would be both wasteful and highly polluting. Applications systems to date have been designed on the basis that lubricant passes once through the mill and is thereafter totally lost.

Figure 13.8 is a sketch of a typical oil application layout for a hot steel strip mill. Oil from the storage tank is metered at a controlled rate by means of

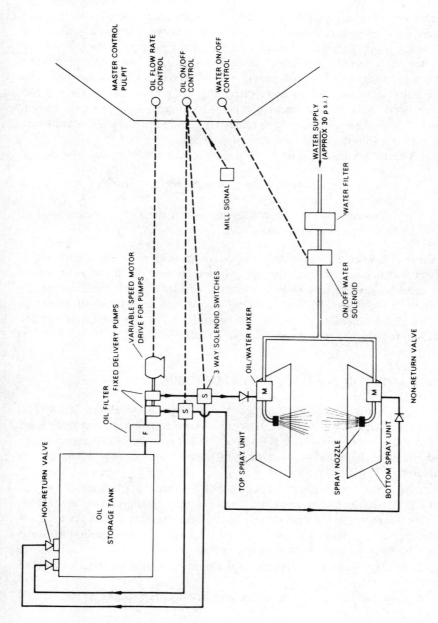

Figure 13.8 Oil-delivery system for the application of hot roll oil

fixed delivery pumps. The oil must only be applied to the mill when steel stock is present, otherwise there will be wastage of oil and mill control difficulties. The requirement then is for an intermittent oil supply synchronised with the steel production flow and, to achieve this, stock-sensing devices are used to provide a mill signal which operates solenoid switches in the oil-supply lines. When the switches are activated by the signal the oil flows are directed to the mill rolls but at all other times the solenoids return the oil flows to the storage tank. The spray units are positioned so as to direct the spraying nozzles at the top and bottom work rolls or their back-up rolls. A continuous flow of water is maintained through these nozzles and the oil feed is arranged to allow the intermittent supply to mix into the water flow prior to spraying.

An effective lubrication system will aim at meeting the following principles:

(1) Lubricant must be supplied at a carefully controlled rate at the appropriate times only.
(2) Lubricant must be applied evenly across the full working widths of both top and bottom work rolls.

These principles are particularly important in tandem finishing mills which rely on friction to drive stock through, without creating the inter-stand tensions that cause serious difficulties in mill control and also impair the shape of the products. Since the application of oil reduces friction the amount of oil has to be carefully balanced to maintain mill control on the one hand, but provide sufficient lubricant to reduce roll wear on the other hand.

LUBRICANTS FOR THE HOT ROLLING OF STEEL

Water is needed in copious quantities as a coolant for the work rolls in order to stabilise the shape of the products. It does provide a certain degree of lubrication for the mill but its properties in this respect are not regarded as being particularly good and it is therefore not surprising that severe roll wear problems are encountered.

Conventional hydrocarbon lubricants can be used to good effect on hot steel mills. Although hydrocarbons are rapidly degraded and pyrolised at 1000 °C, the time taken for the lubricant to pass through the mill in contact with the white hot steel is sufficiently short for it to do an effective lubrication job. Some of the lubricant almost certainly emerges from the exit side of the mill before degradation temperatures are reached, since burning oil is not normally observed on the entry side.

The properties that appear to be desirable in a hot steel rolling oil are:

(1) Good anti-wear and load properties at high temperatures.
(2) Good metal surface adhesion in the presence of water.

(3) Good metal surface wetting properties in the presence of water.
(4) Easy dispersion in water at ambient water temperatures.
(5) Cleanliness in burning from the hot metal surface.
(6) Non-corrosive properties in the neat oil pumping system.

In most application systems, oil is applied as a dispersion in a water carrier. The oil has to compete with water in wetting the metal surfaces at which it is directed. Good wetting properties are required to establish a lubricant film on the roller surfaces under such adverse conditions. Having adhered to the rollers, the lubricant film must remain intact whilst it is transferred to the roll bite despite being bombarded vigorously with jets of mill-coolant waters. The use of a soluble roll oil is impractical in these circumstances since it would easily be washed away.

Good dispersion and surface adhesion properties will ensure that a lubricant film will reach its target in the roll bite. The oil is then required to reduce roll wear under high-temperature, boundary-lubrication conditions. Several different types of base oils have been tried of mineral, vegetable or synthetic origin, and varying considerably in surface adhesion and lubricity properties.

Another important property required in a hot-rolling oil is cleanliness of combustion. The majority of the oil applied to a mill is destroyed by burning from the hot steel and the products of pyrolysis must be innocuous to the mill operators. Highly refined base oils are therefore becoming the preferred choice, particularly those that can be substantially oxidised to carbon dioxide and water.

LUBRICANT RECOMMENDATIONS

No general specifications have been devised for rolling lubricants as yet. Individual rolling plants tend to provide guidelines for the properties of the roll oils they require and in practice these tend to be similar to each other.

14 Planning for All Lubrication Requirements

M. D. Cox *B.Sc.*
J. S. Faircloth
Esso Petroleum Company Limited

Nowadays most factories and plants realise the importance of investing time and effort to study and maintain an efficient lubrication plan for their works, though some still do not. However, not so long ago the application of lubricants was considered a menial task and was left to the unskilled or untrained. The science surrounding the lubricant at all stages in its progress from the oilfield often deserted the highly refined finished product at its point of use. The planning that prevailed in most other functions of management in the pursuit of efficiency was often noticeably absent in the field of lubrication.

But why does lubrication merit planning and the close attention of management and why was it neglected in the past?

The physical and administrative tasks of obtaining the correct lubricant and applying it to the correct machines in the correct amounts are shown in figure

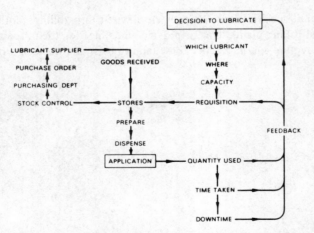

Figure 14.1   Functions of lubrication contributing
to plant-operating expense

14.1. The value of the man-hours represented by each step has been shown to amount to some two to seven times the lubricant cost, the lower figure of twice the lubricant cost being true only when many of the steps in figure 14.1 have been eliminated by the introduction of the centralised lubrication system. A close study of the methods adopted to implement each stage in figure 14.1 in any particular factory presents many opportunities for the plant lubrication engineer to reduce this seven-to-one lubricant ratio by investigation of the following areas:

(1) Number of orders placed with different suppliers.
(2) Number of grades in use.
(3) Method of application, for example lubricars, centralised systems, organised daily work schedules for oilers.
(4) Method of stocking and dispensing.
(5) Number of invoices and statements.

But perhaps the biggest saving will come from the lack of breakdowns as a result of poor lubrication, or where a fault has been reported by an oiler and corrective action has been taken to avoid a major time-consuming failure. Indirect savings are often difficult to calculate but the cost of a production line failure for 24 hours can often be well above the annual lubricants cost.

MANAGEMENT'S RESPONSIBILITY

When each department of a factory takes care of their own interest in lubricants there is often a genuine conflict of interest. Some of the following problems can arise:

(1) Purchasing managers buy lubricants on a price per litre basis only.
(2) Maintenance engineers seek the best and often the most expensive lubricants as recommended by the machine manufacturer. They also require machines to be laid off periodically for routine lubrication.
(3) Production engineers with demanding production schedules resist the shut-down of machines.
(4) Machine operators on bonus rates avoid stopping machines if at all possible.

To avoid this situation developing, either a committee can be formed with representatives from each department or a lubrication engineer can be appointed with an overriding brief. Whichever method is adopted, interdepartmental communication must be maintained.

Lubricant grade selection

Major oil companies produce lubricants to suit the market as they see it, and not as direct equivalents of each other's products. Factories without a detailed knowledge of the suitability of their existing grades to meet new applications can find it impossible to reduce the number of different branded grades they stock. Stocks build up, often based on machine manufacturers' handbook advice accompanying incoming machines and unwittingly containing many different branded products where a small number of carefully selected grades would have been sufficient.

The dangers inherent in using more lubricants than are necessary include:

(1) Purchasing becomes less economic since smaller volumes of each grade are ordered, often necessitating delivery in high-cost packages instead of in bulk. Investment in stock is increased.

(2) Handling costs are increased when a higher proportion of package products are stocked.

(3) The logistics of applying the correct grade to the point of application becomes more complex and time-consuming for the oiler.

(4) Storing, identifying and dispensing requires more highly rated space, equipment and time.

(5) The chances of applying the wrong grade to a point on a machine are increased.

Planned lubrication

How does one organise a planned lubrication programme if it has not been done before? The first step is to appoint a committee, or the lubrication engineer mentioned earlier, to look at lubrication in total. The lubricant supplier can probably assist in this service. However, whether this is done or whether an outside consultant is used, the priority must still be that the plant exists to produce, not just to be lubricated.

The immediate objectives at this stage will be to collect as much information on lubrication methods: (1) in the factory at present, and (2) from research and outside contacts.

Information available in the factory

(1) How much lubricant is used (litres per year) listed by grade name, supplier, how stored, how dispensed and where.

(2) A list of plants giving the following information:

    (a) Machine type, number and location.

    (b) Points to be lubricated.

    (c) Which lubricant to use.

    (d) Method of lubrication.

    (e) Machine capacity.

    (f) Frequency recommended for each lubrication point.

Table 14.1 Lubrication requirements divided into groups

A	B	C	D
Oil nipples	Oil baths	Oil bottles	Mechanical lubricators
Oil cups	Reservoirs	Air-line lubricators	
Grease nipples Grease cups	Sumps	Oil wells	

(g) Machine downtime history under the sub-headings of:
  (i) Total downtime.
  (ii) Downtime directly due to poor lubrication.
  In many cases (g) will be difficult to ascertain.

(3) Manhours spent on lubrication and quality of labour employed at present.

(4) Time standards for each lubrication operation after grouping as in Table 14.1. These groups will be different for different plants. The workload for any machine or workshop can be calculated weekly, monthly or annually as required.

(5) Standards of lubrication design and practice when:
  (a) Purchasing machines — obtain lubricant recommendations in terms of existing factory lubricant suppliers from the machine manufacturer.
  (b) Lubricating elements of plants, such as bearings and electric motors.

(6) Work schedules. A well-designed work schedule will take into consideration:
  (a) Work that can be carried out without interfering with production.
  (b) Work that necessitates machine shut-down.
  (c) Work that can be integrated with routine mechanical or electrical maintenance.
  (d) Maximum productive work by the operators carrying out lubrication.

(7) Pay scales for storemen, oilers, machine operators, etc.

(8) Future factory expansion plans.

Information from research and outside contacts

Systems availability

Most large oil companies are prepared to offer advice and assistance on documentation, equipment, storage, grade rationalisation, and visual control systems, etc. The advantage of approaching oil companies lies in their:

(a) Wide experience of lubrication in all types of industry.
(b) Detailed knowledge of the suitability of their products for all kinds of plant.
(c) Close contact with machine manufacturers.
(d) Eventual involvement as possible suppliers.

Visits to other factories

Visits to factories engaged in a similar operation will show how they tackle the problems and will allow useful practical information to be gained at first hand before plans are finalised.

Equipment manufacturers

Manufacturers offer catalogues and technical advice on a wide range of equipment from grease guns to centralised lubrication systems. The cost and suitability of any equipment to meet particular needs can also be discussed with them.

Quantity discounts for bulk buying

Back to the lubricant suppliers again: it will pay dividends to know their minimum ordering quantities that qualify for concessions so that these factors may be taken into account when deciding on any new storage facilities.

At this stage sufficient information will have been obtained to allow a plan to be formulated. The research information will allow the plan to be as elaborate as required whereas the information from the shop floor will provide the practical limitations that will ensure the plan is based on present and future production needs.

Before presenting a plan that involves expenditure and plant, possible additional staff, reorganising job responsibilities, infringement on bonus schemes and time-off for training, there must be some justification.

Ideally this would be the conversion of the objective into hard cash savings. If factory records are good there is no difficulty but generally there is a lack of historical information. When this is the case an estimate of the present cost should be compared with the possible savings. Basing the approach on the ideal would result in the general format shown in Table 14.2.

There are many items in Table 14.2 where the savings, especially, will be difficult to quantify. However, if it can be agreed that there *will* be savings on any items, the total level of expenditure will give an indication of the possible

Table 14.2   Comparison of total costs with possible
savings after adopting lubrication plan

Cost item	Savings as a result of lubrication planning (£ per year)	Total present expenditure (£ per year)
(a) Spares and maintenance (reduced as a result of planned lubrication)		
(b) Lost production (reduced as a result of planned lubrication)		
(c) Extra power (power requirements may be unnecessarily high if too viscous oils are used and bearings are allowed to run dry)		
(d) Machine life (lubricants prevent wear and extend machine life)		
(e) Lubricants (a similar number of rationalised grades can be bought in a more economic fashion)		
(f) Cost of lubrication (there are many areas in figure 14.1 where costs can be reduced)		
TOTAL		

magnitude of that saving. Management often require to take important decisions on this basis since the cost of obtaining and processing the necessary historical information would be prohibitive or impossible.

Control systems

When the new lubrication programme is agreed by management steps will need to be taken to make sure that the control or feedback is adequate to ensure the plan fulfils any promises made to them. The aims and objectives set when considering lubrication on a planned basis will be instrumental in selecting a control system. All systems should ensure that lubrication proposals are effectively carried out and the degree of capability in producing cost data and daily advice of backlogs or mishaps is the usual variable between them. Ideally the best system will:

(1) allow for the spot interchange of manpower from one section to another, thus being in a position to cope with absenteeism or any form of labour movement between jobs not readily foreseen.

(2) have simple documentation (see figure 14.2).

(3) alert the oiler when major servicing is becoming due.

(4) allow the oiler to determine, before setting out on a routine, the best estimate of the type and quantity of oils and greases required.

(5) reduce to an absolute minimum the possibility of putting the wrong oil into a machine.

(6) allow aims and objectives to be readily measured.

(7) be carried out in normal working hours.

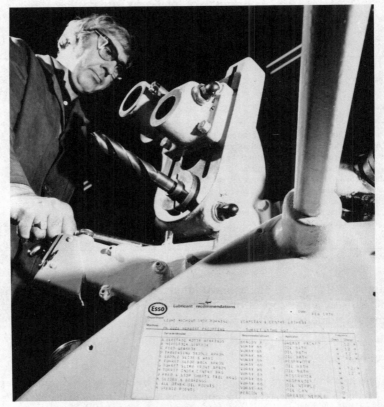

Figure 14.2   A card on each machine tool, showing what parts are to be lubricated at what intervals and with what lubricants, greatly reduces the possibility of a wrong grade being used

Visual system control

If the system does not accommodate feedback of information on action taken at the time of applying the lubricant there is the danger that lubrication points not attended to for legitimate reasons or otherwise will remain unknown to the supervisors and management. These become, of course, weak spots in the production cycle.

This highlights the need for visual aid and figure 14.3 shows system control using pegboard. In this case the filing system on the side of the planning board contains full lubrication details for all machinery to be programmed and colour signals have been allocated frequencies.

Each piece of machinery or machine tool occupies one flow line across the board, which spans a 12 month period. The frequency signals are scheduled on

Figure 14.3   The lubrication systems described in this chapter can be
adapted for use with vehicle fleets. Wall boards of this kind are one
means of keeping track of the necessary actions

this line. Job cards are issued to the oilers each day to cover pre-determined work
schedules. When these are returned they allow the board to be updated. If the job
has been completed the control peg is advanced to the frequency signal for that
date. Each day those responsible for plant lubrication can see by looking to the
right and left of the date:

(1) What the workload is ahead.
(2) What has been completed.
(3) What did not get done.

With this system the work that did not get done cannot be overlooked and calls
for a management decision to bring about a remedy and correct the programme
back to normal.

Computerised lubrication

In a large plant a vast amount of data must be processed to obtain a well-defined
and comprehensive oiler workload for each working day. If a feedback of oil

consumption or a defect-reporting system is superimposed on this workload the task becomes even more complex. The use of computers has therefore been investigated as a method of avoiding much of this routine clerical work.

The present experience of the use of computers in planned lubrication in the UK is as high-speed sorting machines to produce daily workloads and routes for oilers based on an input of:

(1) Machine location, name and number.
(2) Parts to be lubricated.
(3) Products to be used.
(4) Method of lubrication.
(5) Seasonality factor, if applicable.
(6) Frequency of lubrication.
    (a) Check and top up.
    (b) Change.
(7) Time required to perform each task.
(8) Can job be done with machine 'on line' or has it to be shut down?
(9) Tasks already completed.

If the oiler's logical route through the factory is then determined by a personal examination of the shop floor requirements and fed into the computer, it can produce a print-out that itemises the tasks to be accomplished each day in that sequence. It must be remembered that a computer cannot produce a better schedule than can be produced manually and, where data require a lot of manual processing before being fed into the computer, the advantages of using it may be doubtful.

Example of an oiler's route

A typical output from either a computerised or a manual programme could be as follows:

In a factory of 240 machines in an average engineering machine shop, the requirement could be for 1 oiler. The oiler's working month would then be divided into four 5-day weeks. Calling these days 1–20 and calling the machines 101–340 the monthly work programme would be as follows:

Day	Task	Machine No.	Remarks
Day 1	All daily tasks	101–340	Itemised by machine part to be lubricated, type of lubricant and method of lubrication
	All weekly tasks	101–160	
	All monthly tasks	101–115	
	All three-monthly tasks	101–104	
	All six-monthly tasks	101–102	
	All annual tasks	101	

Day	Task	Machine No.	Remarks
Day 2	All daily tasks	101–340	
	All weekly tasks	161–320	
	All monthly tasks	116–130	
	All three-monthly tasks	105–110	
	All six-monthly tasks	103–105	
	All annual tasks	102	

and so on

On the 5th, 10th, 15th and 20th day of each month, major tasks would be performed as well as the daily checks on machines 101 to 340. Therefore day 5 might read:

Daily checks Machines 101–340 and all major oil changes on machines 101–120.

A simplification of the above programme is to class all tasks over monthly as major tasks, which is often the case, and then the workload from one month to another will be exactly the same. The major tasks can then be highlighted on a wall planning chart calendar to ensure that more detailed information is available on the progress of these important areas.

Feedback can be achieved on items requiring repairs by attaching a label to the faulty machine with a tear-off section that outlines the repairs necessary being sent to maintenance department, who fills in the details of any attention given.

This chapter is not intended to be a formula for instant planned lubrication. Its purpose is to highlight the importance of planning lubrication and the cost of neglecting it, and to outline the steps that need to be taken to ensure that each machine receives the correct lubricant at the correct time in the most efficient manner possible.

15 Lubricant Storage and Handling

M. D. Cox *B.Sc.*
M. P. Redgard
Esso Petroleum Company Limited

The chief requirement when handling lubricants is to avoid contamination by dust, grit and water. Many applications require specialised lubricants but all the care with which a lubricant has been formulated, manufactured and packaged can be nullified by the presence of small quantities of contaminant.

The maintenance of a clean and well-designed oil store and a high standard of dispensing and lubricant storage is therefore important to:

(1) prevent contamination from dirt or water
(2) prevent confusion of brands by defaced markings
(3) encourage responsible attitudes to machinery maintenance and lubrication
(4) ensure correct storage conditions for temperature-sensitive lubricants
(5) prevent accidents that could arise from leakage, and eliminate health hazards
(6) prevent fires that could arise from spills, leaks and from bad housekeeping
(7) save time, money and promote efficiency
(8) ensure adequate stock levels and ease of application
(9) ensure that rigid quality control by oil companies is not nullified
(10) reduce down-time and so increase machine output

BLENDING AND PACKAGING

Blending

To appreciate the need for a high standard of storage and handling one first needs to look at the standards maintained by the large oil companies and the requirements of industry. From the time the oil arrives at the refinery until it is put into packages or held in bulk at the blending plant its purity and quality are constantly being monitored in the laboratory. Manufacturers of modern plant and machinery are often designing to the known limits of oil technology and they demand lubricants that meet very high standards of performance and purity.

Packaging

In industrial lubrication there are two main package types for oil and four for greases.

Oil
Barrel: 205 litres
Non-returnable drums: 25 litres

Grease
Barrel: 180 kilograms
Keg: 50 kilograms
Pail: 12.5 kilograms
Tin: 3 kilograms

Only the oil and grease barrels are returnable and consumers are charged a deposit on these containers which is refundable if the barrels are returned in good condition. No matter how good the apparent condition of the returned barrels, they are cleaned, using special equipment, and reconditioned by removing dents and re-rolling the seams. After shot blasting they are tested for leaks and given a final rinse before being dried and visually inspected. Any residue is removed by vacuum extraction. The bungs that are inserted before painting are removed again at the filling point, where a final inspection is made. With this method of inspection a barrel may safely make several 'trips' before being discarded.

Bulk deliveries

The alternative to packages is bulk deliveries and there are many advantages in this method of receiving products: the operation requires very little labour; the delivery takes a shorter time to complete; there are no charges for returnable packages; and empty containers are avoided, saving space, loss as a result of damage, and loss as a result of clingage in the empty barrel (this can be up to 5 litres in the case of viscous oils).

Bulk is therefore the preferred storage method, and should always be considered by anyone using more than 3500 litres a year of any particular grade.

ORDERING, DELIVERING AND RECEIVING

Ordering

A typical sequence of events in a well-organised store would be:

(1) A lubricant X is withdrawn from the store by maintenance department to meet the normal requirements of plant lubrication schedules.

(2) The storeman notes from his records that this withdrawal reduces the store stock of grade X below re-order level. (This level would be predetermined at from two to three times the volume of oil X used during the normal period taken for the oil company to deliver from the time the storeman places his requisition on his own purchase department.)

(3) The storeman raises a requisition to his purchase department for grade X and any other grades that are approaching re-order level. (The fewer requisitions and purchase orders placed the less paper generated all round and for this reason it is often advantageous for the storeman to order direct from the oil company.)

(4) The purchase department then places the complete order with the oil company.

Delivery

An order for bulk and packaged lubricants will be delivered separately, as specialised trucks are used in each case. The package truck has a tail-gate hoist for loading and unloading heavy barrels, whereas bulk deliveries are made in compartmented bulk tankers. However, in each case the delivery will be accompanied by an advice note.

Receiving

All incoming deliveries should be checked by the receiving company as follows:

(1) The advice note should be checked against the purchase order and the delivery to ensure that the correct order has been received.

(2) All package seals should be intact and no packages should be leaking.

(3) All grade markings should be legible.

(4) Before deliveries are made to bulk storage the tank and the tanker should be double checked to ensure that the correct oil is received into the correct tank in the correct quantity (grade cross-overs and over-filling of tanks cause machine down-time, accidents and pollution).

(5) Bulk oil tankers are heavy (30–40 tonnes) and hard standing should be provided beside the fill pipes to prevent subsidence or damage to roadways.

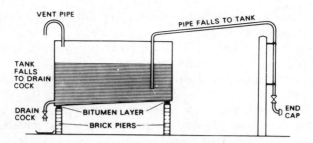

Figure 15.1 Bulk oil tank layout

(6) All fill pipes and outlet pipes should be fitted with stop valves and dust caps and clearly marked with the grade to be stored. Each storage tank should be fitted with a vent pipe and a calibrated dip stick or a contents gauge (see figure 15.1).

(7) 205-litre barrels of oil or grease should never be dropped from vehicles as personal injury, barrel damage or leakage might well result. If a fork lift or tail-gate hoist is not available a barrel skid should be made available to unload these items.

STORE DESIGN

An existing factory or plant will have some area designated for lubricant storage and it may not be possible to change either the layout or the location. However, frequently this is possible, and the following points will assist when designing a new store or re-designing an existing one.

Outdoor storage

Owing to the very high capital and rated cost of industrial buildings, barrels may have to be stored outside. When this is the case, observation of certain basic principles minimise the risk of contamination:

(1) Barrels should be stored on their side, or tilted so that moisture cannot be breathed through the access bungs.

(2) Barrels should be stored off the ground on battens or preferably on an oil-resistant base or barrel rack. They should never be left lying on bare ground.

(3) Barrels should be stored with both end bungs below the internal liquid level to keep the seals moist and airtight.

(4) If stored on their side one on top of the other the end barrels should be securely chocked to prevent rolling.

(5) A first-in, first-out, policy should be adopted even though this may be difficult owing to the method of stacking.

(6) The store should not be situated in a dusty or corrosive atmospheric environment.

(7) A roof is desirable if at all possible and some form of cover is essential for all packages smaller than barrels.

(8) Stocks should be kept to levels that will allow a reasonably quick turnover.

(9) Some water-based oil emulsions are sensitive to high and low temperature and other products containing solvents need protection from the sun or direct heat. They should be preferably stored indoors.

(10) If heating is necessary to assist the passage of product from bulk storage the heat flux of the heater should not exceed 5 kW/m^2 for heavy EP oils and 23 kW/m^2 for light straight mineral oils, and the bulk oil temperature should not be allowed to exceed 70 °C otherwise cracking of the oil and/or degradation of the additives may occur.

Indoor storage

The desirable features in an indoor store are:

(1) Easy access for lubrication trolleys and fork lift trucks.

(2) Ability to operate a first-in, first-out, policy.

(3) Separate areas for stocks being used and new stocks.

(4) Good ventilation.

(5) Light, clean walls.

(6) Smooth level oil-resistant floor. The whole store should be built of fireproof materials.

(7) Heating is useful for keeping viscous oils fluid. It will certainly be required if a storeman works there permanently.

(8) Depending on the factory demand, bulk storage and barrel racking laid out in an orderly fashion.

(9) Ability to lock the store, which should be completely partitioned off from the building in which it is housed.

(10) Ability to supervise and record receiving and issuing of stocks.

OIL STORAGE EQUIPMENT

This can be broken down into handling, storing and dispensing equipment.

Handling equipment

This is essentially used for removing heavy barrels and ranges from the versatile fork lift truck (figure 15.2) to the humble barrel skid.

Figure 15.2   The fork lift truck is one of the most versatile
pieces of equipment for handling heavy barrels
(courtesy of Eaton Ltd)

Storing and dispensing

The most convenient method of storage is the bulk tank which involves the minimum of handling (figure 15.3). However, barrels are a practical alternative for small throughput grades if they are handled with care. Proprietary racking, the girders below bulk storage tanks and barrel tilts are used in conjunction with barrel taps and drip trays as a convenient method of oil storage and dispensing. If there is a fear that taps will be left on, presenting a spillage hazard, semi-rotary or barrel pumps can be inserted into the upright barrels to avoid this problem when oil is dispensed indoors.

Grease dispensers are a worthwhile investment whenever sizeable volumes of grease are applied by hand guns.

For cutting oils, accurate water to oil ratios can be achieved by using a Fluidmix valve (figure 15.4).

Figure 15.3   A typical bulk lubrication store

TANKS AND PIPEWORK

Tanks are usually fabricated of mild steel plate and are available in many different standard sizes. However, if heavy heaters or pumps are attached to tanks they may require to be built of heavier than normal gauge metal.

If protection against driving rain and snow is provided the tanks can be sited outside but the first choice would always be indoors. In the majority of cases the best site for bulk tanks is in the area used for open packages. As delivery tankers can pump oil against a head of up to 10 m, bulk tanks are often sited well above ground to take advantage of gravity for the outflow.

In designing pipework and tank storage systems oil viscosity should ideally be less than 2500 cSt for easy reliable gravity flow out of the tank and 1000 cSt for easy pumping round a ring main. Since viscosity varies from one product to another and also with temperature the following parameters should be considered at the design stage.

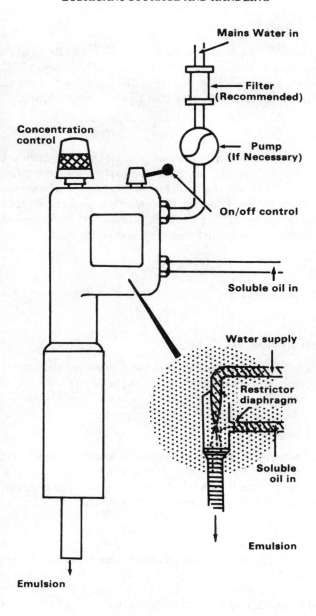

Figure 15.4 The Fluidmix valve brings together water and a soluble oil to form a soluble-oil emulsion. Concentration can be adjusted instantaneously by a knob at the top of the housing (courtesy of Ronald Trist Controls Ltd)

Ambient temperature

In the UK, the contents of the tanks situated in a workshop, where day temperatures require to be maintained at 13 °C, will rarely fall below 2 °C. The same tank contents in outside storage will rarely fall below −4 °C even although the ambient temperature drops well below that figure for short periods.

Pipe diameter

In gravity or pumped systems the higher the oil viscosity, intrinsically or as a result of low temperature, the slower the flow rate for a given pump or gravity head. Oils of 2500 cSt will require a high pump pressure and a minimum pipe diameter of about 40 mm. Preferably the viscosity should be reduced to 1000 cSt for pumped systems by heating the storage tanks. The pipe diameter will be a cost compromise between pump horsepower and piping costs.

Flow rate

The demand determines the gravity or pump head required for a given oil viscosity.

Pump h.p./gravity head

This requires to be increased as the viscosity is increased and the pipe diameter is decreased. It is also dependent on the length of pipework involved and the flow rate required.

Pipe and tank lagging

50 mm thick covering of 85 per cent magnesia lagging can reduce tank heat loss by up to 75 per cent. The cost of lagging should be assessed in relation to the heating cost. For extremely viscous oils, exposed pipes should also be lagged and in some rare extreme cases traced electrically as well.

Tank heating

This can be performed by electric immersion heater or steam coil. Important parameters are:

(1) The heat flux of the heater should not exceed 5 kW/m² for heavy EP oils and 23 kW/m² for light straight mineral oils.
(2) The steam temperature of the coil should not exceed 120 °C.
(3) The bulk oil temperatures should not exceed 70 °C. Outflow heaters are rarely an economic proposition because of the high surface area and high rating required for even low flow rates.

In the normal design situation ambient temperature, flow rate, pipe length and oil viscosity will be fixed by the factory location, lubricant demands, geography and application respectively, whereas the gravity or pump head, the pipe diameter, heater size and lagging will be matters for assessment depending on cost and suitability.

The latter design variables should be examined by considering the following alternatives in order:

(1) Increasing the pipe size using gravity flow.
(2) Increasing the pipe size and including a pump in the system.
(3) Increasing the pipe size, introducing a pump and also an immersion heater.

STOREKEEPING – INVENTORIES AND STOCK ROTATION

There are several simple symbol systems available that are used to identify lubricant type and frequency of application. BS 4412:1969 and PERA (Production Engineering Research Association) systems are similar but should not be confused with the German DIN 51502:1979 specification which attributes different meaning to the frequency shapes.

Having agreed on a uniform identification system throughout the factory, stock levels for each grade should be established. This will be a compromise between:

(1) The order quantity that qualifies for the best delivery discount.
(2) The annual factory consumption.
(3) The oil consumption during the period from the time a requisition is placed on purchase department until the oil company delivers the order.
(4) Seasonal peaks in demand, for example, major oil changes or overhaul periods.
(5) Contingency factor for unexpected leakage from large capacity systems.
(6) Future factory expansion plans.

A minimum tank size would be 2 months' average usage, and frequently twice this size of tank is installed.

It is essential to keep a record of all oil received and issued to the factory so that a paper tally of stocks is always available to compare with actual stocks for ordering purposes. Large differences will be due to leakage or pilfering if book-keeping is correct.

Re-order levels can also be established from a consideration of the above points.

Having decided the stock level, provision must be made for storing and dispensing. Whatever the size of the store the oil being used should be kept separate from the reserve being stored and a system introduced to use the oldest packaged stocks first.

Storage provision for handling equipment, measures, overalls and oily rags should be made. This encourages a systematic approach to lubrication whilst lessening the risk of fires and accidents.

Empty containers

Empty containers should have their bungs replaced before being taken away from the store. Returnable items should be stored near the receiving area so that thay may be returned to the supplier as soon as possible. This allows credit to be received quickly and minimises the risk of unauthorised removal or damage.

LEAKAGE AND CONTAMINATION

Drip trays below all taps are essential. Spilt oil is a fire hazard and can cause accidents: it should be wiped up immediately with an absorbent material. Although sawdust is often used for this purpose it is now considered bad practice because of fire risk; there are equally effective proprietary crystalline materials available. However, where sawdust is the only material available for this purpose it should be lifted as soon as it becomes oily and disposed of into a bin with a lid. There are several ways in which contamination risks can be lessened:

(1) Easy access to fill pipes and draw-off cocks.

(2) Clear and legible labelling of fill pipes and draw-off cocks.

(3) Each container used in the works should be marked with the grade it is intended to carry and should be used only for that grade.

(4) When new packages are opened they should be compared with previous stocks for colour and consistency and set aside for return to the supplier if there is any variation from normal.

(5) Area around bungs and lids should be wiped clean before opening containers.

(6) Spare containers should be kept on shelves to prevent grit or dirt getting into them.

(7) Special care must be taken with grease which should be dispensed from special equipment if possible. If other methods, for example, spatulas or hands, are used to fill guns, the grease should be bought in minimum size containers. Also the lids of these containers should not be laid on the floor while grease guns are being fitted as this is another source of contamination.

(8) Drains in and near the oil storage area should be fitted with oil interceptors.

LUBRICATING AND DISPENSING EQUIPMENT

The choice of lubricating and dispensing equipment depends on the result of an investigation into the planning of lubrication duties in the plant (chapter 14). However, the potential choice ranges from an oil can with the grade name clearly marked on it to a fully automatic centralised lubrication system. This range is outlined below.

Oil cans and hand grease guns

These simple pieces of equipment are often required to reach inaccessible points on equipment.

The original supply container

This system has the merit of avoiding many of the normal contamination and cross-over risks associated with a decanting into an intermediate vessel, but is often very labour intensive if a lifting device is not available.

Figure 15.5   A lubrication trolley with its own dispensing pumps

Lubricar trolleys

Although these items (figure 15.5) are frequently introduced into a factory lubrication schedule they are often discarded by the oiler in favour of much simpler equipment because they have been designed without due consideration of the available space between machines. However, if properly designed they can be an alternative to re-locating the store in a more central position or opening up sub-stores in outlying parts of the factory.

Waste oil units

A suction pump and tank is required to remove certain cutting oils and sump oils. Sometimes an upright barrel on a trolley can be fitted with ancillary pumps and hoses necessary to perform this function. Proprietary devices are also available that can be fitted with a filter to allow the clean oil to be returned to the sump for further use.

Centralised lubrication system

Oil and greases can be pumped along pipework and ring (figure 15.6) mains for the following purposes:

Figure 15.6   A range of lubricants being dispensed in
a vehicle maintenance workshop

(1) To lubricate, under pressure, gears and bearings in large plants such as steel mills.

(2) To circulate cutting oils to central draw-off points for several machines or to the actual machines themselves. In the latter case the oil, after passing over the workpiece, returns to the storage tank where it is filtered before being pumped back round the system.

(3) The lubrication of a large number of grease-filled bearings on a machine can often be most easily achieved by feeding grease from a central container through special control valves that meter out pre-determined quantities. Pop out indicators locate malfunction or grease starvation of any of the metering valves.

Drain period

These can vary from 'filled for life' to 'once through' systems and the advice of the oil supplier and machine manufacturer is vital in this case. The main criteria for oil life is the degree of severity of environment and operation and this will vary from factory to factory dependent on the following conditions:

(1) Atmospheric environment, for example dusty or corrosive atmosphere.

(2) Standard of housekeeping, for example contamination due to the wrong grade of oil being used or from the use of dirty containers.

(3) Moisture in the atmosphere, since most systems breathe air.

(4) High operating temperature – this causes oxidation of the oil, degradation of heat-sensitive additives and in extreme cases cracking of the oil.

(5) Self-contamination caused by bearing particles from the running-in process, the overloading of machinery, or leakage – for example leakage of cutting oils into a machine tool bearing.

SPECIAL CARE GRADES

Certain oils require special care in storage and handling. Some of these oils are listed below.

Electrical oils

These oils are refined to very close tolerances and have a very low water content. They are transported in special stainless steel tanks used only for this purpose. If contaminated with metal salts or water they will lose their insulating quality and permit arcing while in service; this could cause fires or explosions.

Turbine oils

Cleanliness is also the primary requirement here as turbine oils are frequently expected to last 10–20 years. Centrifuges on the turbines ensure any steam condensate and solids can be removed. Other contaminants, for example small quantities of engine oil, can affect their water-separating properties to such an extent that the oil may have to be discarded. This can be expensive as these are high-cost quality products and systems tend to be large.

Hydraulic oils

Although these oils are often the most sophisticated in common use and are designed to cope with the inevitable contamination from atmospheric moisture arising from high- and low-temperature operation, they cannot protect the system against grit or dirt. Dirt is the great enemy of this low clearance system and great care must be observed when storing and dispensing. A filter gauze should always be used when re-charging reservoirs and in-line filters should be examined often. Engine oil contamination even in minute amounts can destroy the water-separating properties of this oil.

Cutting oils

Neat cutting oils contain special ingredients to help them perform one of the most arduous lubrication tasks in industry. It is especially important that operators do not wear clothing that has been soaked in cutting oil as this can cause skin irritation if continued for long periods. The risk of skin damage and dermatitis is increased if the oil is contaminated with metal or other debris. Metal chips, swarf and grinding wheel particles are also detrimental to the oil additives and the moving parts of the coolant system, beside imparting a poorer finish to the workpiece and shorter life to the tool.

Soluble oils are sensitive to both high (above 40 °C) and low (below 0 °C) temperature and should be stored inside if at all possible to prevent separation of the water by freezing or evaporation. Soluble oils are incompatible with most other oil types and often with other soluble oils.

Neat cutting oils are less sensitive to temperature but can suffer separation of the fatty content in extreme cold. Contamination with other oils can also cause scaling and poor finish.

Greases

Greases are often the most liable to contamination of all oil products. This is because they are solid and cannot be dispensed without mechanical means, usually either a spatula, hand or some proprietary gun filling device. Unless surgical-like cleanliness is observed with the first two methods of dispensing, contamination is almost inevitable and therefore the smallest available container should be stocked. Contamination also occurs at the point of application. Grease nipples with a residue of grease from the previous service attract airborne grit which can be pumped straight into the bearing if it is not removed before applying the grease gun. Overfilling, especially of electric motors, is so common that some factories do not include such equipment on the lubrication schedule, but prefer to re-pack them annually when major overhauls become due. Centralised systems are particularly suitable for grease, the advantages being:

(1) Metered lubrication of required volume of grease to all points regularly.
(2) Contamination is reduced to a minimum.
(3) Any fault in the system is displayed visually.
(4) Labour costs are almost eliminated.

Grease is not sensitive to temperature but cold grease should not be connected into a central system before being allowed to reach its working temperature.

Index